Instructor's Guide with Solutions for

THE BASIC PRACTICE
OF STATISTICS

David S. Moore
Purdue University

Darryl K. Nestor
Bluffton College

W. H. FREEMAN AND COMPANY
New York

ISBN 0-7167-2676-9

Printed in the United States of America

Second printing 1995, RRD

CONTENTS

SOLUTIONS TO EXERCISES

CONTENTS

INTRODUCTION

This Instructor's Guide tries to make it easier to teach from *The Basic Practice of Statistics* (BPS). The most important part is without doubt the full solutions of all exercises prepared by Professor Darryl Nestor of Bluffton College, which make up the second part of the Guide. The first part of the Guide contains some specific teaching helps, such as additional examples and new data sets for class use, sample examinations, and suggestions for using the *Against All Odds* and *Decisions Through Data* videos in class. It also contains brief discussions in which I explain why I chose the approach taken in BPS for presenting statistics to beginning students. Reasonable people can (and do) differ about the nature of basic instruction in statistics. My comments in this Guide are one side of a many-sided conversation. Your position in this conversation will no doubt differ at some points. I hope I have at least made clear the reasons for my choices.

I welcome comments, suggestions for improvement of BPS, and reports of errors that escaped detection and can be fixed in new printings. You can reach me at:

Department of Statistics
Purdue University
West Lafayette, IN 47907-1399
Telephone: (317) 494-6050
Fax: (317) 494-0558
email: dsm@stat.purdue.edu

David S. Moore

1 TO THE INSTRUCTOR

1.1 Introduction

The Basic Practice of Statistics (BPS) is an introductory text in statistical ideas and methods for students with limited mathematical background. The Preface in the text attempts to explain and in part to justify the approach taken, and I will not repeat that material here. When compared with most texts for beginners, BPS is characterized by:

- It is relatively short. The essential ideas of a first course are in place after Chapter 6. From that point, we are acquainting students with specific methods that may (or may not) be useful to them. There is adequate material in Chapters 7 to 10 to offer some selection at the end of a one-semester course, but I have avoided encyclopedic coverage in favor of a less intimidating and less expensive book. BPS presents inference for two-way tables, one-way ANOVA, and simple linear regression. It does not cover two-way ANOVA, multiple regression, process control (except for a brief introduction in Chapter 4) or nonparametric procedures (unless the Pearson chi-square test counts as such).

- There is more attention to data analysis. Chapters 1 and 2 give quite full coverage. It is now becoming common to emphasize data in a first course, but many texts still begin with a quite brief treatment of "descriptive statistics."

- There is more attention to designing data production. It is surprising to a practicing statistician how little attention these ideas, among the most influential aspects of statistics, receive in most first courses. Chapter 3 discusses sampling and experimental design, with attention to some of the practical issues involved.

- Probability gets less attention. The Preface explains why. I avoid emphasis on the laws of general probability, and instead stress *distributions*. The presentation is arranged to motivate the study of probability distributions from distributions of data and from the variability of the results of random sampling. The core treatment is quite brief, and I recommend trying it. If you assign the optional sections 4.2 and 4.4, students will meet all the usual material on distributions, but some familiar content on general probability laws is omitted.

- There is more discussion of the ideas of inference. Chapter 5 (with the introduction to sampling distributions in Chapter 4) is the core of the presentation of inference. The ideas aren't easy, but are the key to an understanding that is more than mechanical.

- The presentation of significance tests emphasizes P-values rather than probabilities of Type I and Type II errors and tests with fixed α. This reflects common practice and helps students understand the output of statistical software. The alternative approach appears in an optional section.

- There is more attention to statistics in practice. Realism may be too much to claim in a book that is genuinely elementary. Nonetheless, Chapter 3 describes the practical difficulties of producing good data and the exposition and examples in Chapters 6 to 10 raise many issues that arise in applying inference methods to real problems.

Upon completion of a course based on BPS, students should be able to think critically about data, to select and use graphical and numerical summaries, to apply standard statistical inference procedures, and to draw conclusions from such analyses. They are ready for more specialized statistics courses (such as applied regression or quality control), for "research methods" courses in many fields of study, and for projects, reports, or employment that require basic data analysis.

1.2   Calculators and computers

The practice of statistics requires a good deal of graphing and numerical calculation. Doing some graphing and calculating "by hand" may build understanding of methods. On the other hand, graphics and calculations are always automated in statistical practice. Moreover, struggling with computational aspects of a procedure often interferes with a full understanding of the concepts. Students are easily frustrated by their inability to complete problems correctly. Automating the arithmetic greatly improves their ability to complete problems. I therefore favor automating calculations and graphics as much as your resources and setting allow.

All students should have a calculator that does "two-variable statistics," that is, that calculates not only $\overline{x}$ and s but the correlation r and the least-squares regression line from keyed-in data. BPS is written so that a student with such a calculator will not often be frustrated by the required calculations. Even if you use computer software, students should have a calculator for use at home and on exams. Two-variable statistics calculators sell for less than $20 in 1994. I don't discuss anachronistic "computing formulas" that presuppose a four-function calculator.

Some scientific calculators offer "tables" of standard distributions and some statistical inference procedures. Graphing calculators now provide selected statistical graphs such as histograms and scatterplots. Calculators have the great advantage that students own them, carry them around, and take them home. If everyone in the classroom has a graphing calculator, class discussions can take on new dimensions: pose a problem and let everyone work on it. Students who took advanced math in high school are often familiar with graphing calculators when they arrive in our classes. If your circumstances favor use of a specific type of graphing calculator, by all means do it.

Unfortunately, graphing calculators are better at mathematics than they are at statistics. Data input and editing are cumbersome, making it difficult to do data-oriented problems in class. I'm not impressed by data graphics on tiny screens. Graphing calculators are at least as hard (I think harder) for inexperienced students to learn to use effectively as Windows or Macintosh menu-driven statistical computer software. Like software, course-wide use of a graphing calculator will require devoting some class time to instruction about the technology. If computers are easily accessible, I recommend jumping all the way to software, bypassing fancy calculators.

Computer software—even a spreadsheet—will automate graphics as well as calculations. Use it if you can. Graphics are very important to a modern presentation of basic statistics, so avoid small "student" programs with primitive graphics. The student versions of professionally written software, such as Mystat and the student version of Minitab, are very competent and quite inexpensive. BPS doesn't presuppose use of software, let alone any specific package. There is a separate Minitab guide keyed to BPS for those who use that common software package. A good deal of computer output in appears BPS, from four different packages. (They are Data Desk, Minitab, SAS, and S-PLUS. These programs have quite distinct strengths, and I intend no endorsement by including output from them.) The output appears because any statistics student should become accustomed to looking at computer output and should be able to recognize terms and results familiar from her study.

1.3 The data disk

If you make use of software, you will want to make available the data used in the text. A data disk for *The Basic Practice of Statistics* is available in both DOS and Macintosh formats. The disk contains all data sets with 10 or more entries. The disk also includes a few data sets for which only a graph or output from statistical software appears in the text. These are:

- Exercise 1.33: Calories and sodium in hot dogs, by type.

- Exercise 1.70: Monthly returns on Wal-Mart common stock for 228 consecutive months.

- Figures 4.1 and 4.2: Sample counts of successes for SRSs of sizes 100 and 2500 from a population with $p = 0.6$.

In addition, the data disk contains data files for the examples given in the CHAPTER COMMENTS sections of this Guide and in the sample examinations.

There is a separate file for each data set. The files are named according to the following convention:

tab1-1.dat	Table 1.1
ex1-10.dat	Exercise 1.10
em1-7.dat	Example 1.7
guide1.dat	Instructor's Guide data set 1

A data set that is used several times in a chapter appears under the location where it is first used. Data sets used in several chapters appear several times on the disk to eliminate the need for back references. To allow reading by any software, all data files are in plain text (ASCII) form, with only numerical entries. This requires that character entries be coded as numbers. The coding is usually obvious from comparing the data file with the data table in the text, but the notes in the README file on the disk make the coding explicit.

1.4   Using video

One of the most effective ways to convince your students that statistics is useful is to show them real people (not professors) employing statistics in a variety of settings. Video allows you to do this in the classroom. Two related video series that contain many short documentaries of statistics in use "on location" are:

- *Against All Odds: Inside Statistics.* This telecourse, consisting of 26 half-hour programs, was prepared by COMAP for the Annenberg/Corporation for Public Broadcasting Project. It is available in the U.S. at the subsidized price of $350. Call 1-800-LEARNER for information or to order a copy.

- *Statistics: Decisions Through Data.* This set of 21 shorter modules (5 hours total) is intended for use as a classroom supplement in secondary schools. It was prepared by COMAP for the National Science Foundation and draws on the location segments of *Against All Odds.* It is available for $350 from COMAP. Call 1-800-77-COMAP for information.

If you are outside the United States, you can obtain information about both video series from

COMAP Inc.
Suite 210
57 Bedford Street
Lexington, MA 02173 USA
Fax 1-617-863-1202

Because I was the content developer for these video series, they fit the style and sequence of *The Basic Practice of Statistics* well. In several cases, data from the videos appear in the text. Nevertheless, I do not recommend showing complete programs from *Against All Odds* in the classroom. The shorter modules from *Decisions Through Data* are more suitable for classroom use, but I recommend only a few of them. Video is a poor medium for exposition, and it leaves viewers passive. It is therefore not a good substitute for a live teacher.

Nonetheless, video has several strengths that make it an ideal supplement to your own teaching. Television can bring real users of statistics and their settings into the classroom. And psychologists find that television communicates emotionally rather than rationally, so that it is a vehicle for changing attitudes. One of our goals in teaching basic statistics is to change students' attitudes about the subject. Because video helps do this, consider showing video segments regularly even if you don't think they help students learn the specific topic of that class period. You can find more discussion of the uses of video, and references, in David S. Moore, "The place of video in new styles of teaching and learning statistics," *The American Statistician* 47, 1993, pp. 172–176. This article contains most of what I know about using video to teach.

Here are some specific suggestions for excerpts from *Against All Odds* (AAO) and *Decisions Through Data* (DDD) that work effectively in class. The comments in each case state what section of *The Basic Practice of Statistics* (BPS) the video illustrates.

- *What is Statistics?* is a collage of excerpts from later stories that makes up the first half of Program 1 in AAO and forms the first module of DDD. You can order an edited 14-minute version under this title from the American Statistical Association. I show this in the first class period, and I recommend it even if you use no other video material. (Introduction.)

- *Lightning research* from Program 2 of AAO, Module 3 of DDD. A study of lightning in Colorado discovers interesting facts from a histogram. (Section 1.1. Figure 1.3 in BPS comes from this study.)

- *Calories in hot dogs* from Program 3 of AAO and Module 5 of DDD. The five-number summary and box plots compare beef, meat, and poultry hot dogs. (Section 1.2. Exercise 1.33 in BPS concerns these data and Exercise 2.9 is based on the same article from *Consumer Reports*.)

- *The Boston Beanstalk Club* from Program 4 of AAO, Module 7 of DDD. This social club for tall people leads to discussion of the 68–95–99.7 rule for normal distributions. (Section 1.3.)

- *Saving the manatees* from Program 8 of AAO, Module 11 of DDD. There is a strong linear relation between the number of power boats registered in Florida and the number of manatees killed by boats. (The manatees can illustrate any of Sections 2.1 (scatterplots), 2.3 (least-squares regression), or Chapter 10 (regression inference) in BPS. Updated data appear in Exercise 2.4, and Exercises 2.5, 2.9, and 2.85 also concern these data. The data appear again in Exercise 10.10, and Exercise 10.11 also uses them.)

- *Obesity and metabolism* from Program 8 of AAO and Module 12 of DDD looks at the linear relationship between lean body mass and metabolic rate in the context of a study of obesity. (Sections 2.1, 2.2, and 2.3. Data appear in Exercise 2.7 and are also used for Exercise 2.23.)

- *Sampling at Frito-Lay* from Program 13 of AAO or Module 17 of DDD illustrates the many uses of sampling in the context of making and selling potato chips. (Section 3.1.)

- *The Physicians' Health Study* from Program 12 of AAO (Module 15 in DDD) is a major clinical trial (aspirin and heart attacks) that introduces design of experiments. (Section 3.2. Data from the Physicians' Health study appear in Exercise 2.87 and Exercise 3.49 asks a simplified version of the design.)

At this point, AAO and DDD diverge. Because *Decisions Through Data* is intended for use in secondary schools, its final hour consists of three longer modules (about 20 minutes each) on the fundamental principles of statistical inference. The use of location shooting and animated computer graphics in these units can't be duplicated in the classroom; these are the only three units from either AAO or DDD that I recommend showing in their entirety. They complement Chapters 4 and 5 of BPS. *Against All Odds* continues with coverage of many of the specific methods of inference from Chapters 6 through 10 of BPS.

- *Sampling distributions* are perhaps the single most important idea for student understanding of inference. Module 19 of DDD presents the general idea, the basic facts about the sampling distribution of the sample mean $\overline{x}$, and the application of these ideas to an $\overline{x}$ control chart. The setting is a highly automated AT&T electronics factory. (Sections 4.1, 4.5, and a bit of 4.6.)

- *Battery lifetimes* from Program 19 of AAO lead to an animated graphic that illustrates the behavior of confidence intervals in repeated sampling. Module 20 in DDD is a presentation of the reasoning of confidence intervals using the same setting that can be shown in its entirety. (Section 5.1.)

- *Taste testing of colas* is the setting for an exposition of the reasoning of significance tests in Module 21 of DDD. I prefer this treatment to that in AAO. (Section 5.2 uses the same example to introduce tests; Example 6.2 in the following chapter applies the t procedures to the cola data.)

- *Welfare reform* in Baltimore, from Program 22 of AAO, is an comparative study of new versus existing welfare systems that leads to a two-sample comparison of means. (Section 6.2. Exercise 6.60 concerns some data from this study.)

- *The Salem witchcraft trials*, revisited in Program 23 of AAO, show social and economic differences between accused and accusers via comparison of proportions. (Section 7.2.)

- *Medical practice*: Does the treatment women receive from doctors vary with age? This story in Program 24 of AAO produces a two-way table of counts. (Chapter 8.)

- *The Hubble constant* relates velocity to distance among extra-galactic objects and is a key to assessing the age of the expanding universe. A story in Program 25 of AAO uses the attempt to estimate the Hubble constant to introduce inference about the slope of a regression line. (Chapter 10. Figure 2.11 is a scatterplot of Hubble's data, which are used in Examples 2.11 and 2.12 to illustrate some descriptive facts about regression and correlation. Exercise 2.40 concerns this scatterplot.)

2 PLANNING A COURSE

2.1 Introduction

In preparing to teach from BPS, look carefully at the **Chapter Review sections** that conclude each chapter. There you will find a detailed list of the essential skills that students should gain from study of each chapter. I put these learning objectives at the end of the chapters because they would make little sense to students in advance. You can use them for advance planning as you decide what to emphasize and how much time to devote to each topic.

Also look at the **embedded exercises**, the short sets of exercises that appear frequently within the body of text sections. These cover the specific content of the preceding exposition. Their location tells students "You should be able to do this right now." They also show the instructor what students can be expected to do at each step. The longer sets of **section exercises** at the end of each section ask students to integrate their knowledge, if only because their location doesn't give as clear a hint to the skills required. The **chapter exercises** that follow the chapter review add another level of integration. You can help students by judicious selection of exercises from all three locations.

I have been partially converted by the movement to reform teaching in the mathematical sciences. One of the reformers' emphases is that we should make our classrooms as interactive as possible, involving students in discussion, reaction, problem-solving and the like. Like all who try this, I find that my course outlines cover a bit less material—but that the students master more of it. My outlines reflect this; your mileage may vary. I should add that mature students who have learned how to learn can create their own interaction with text and lecture, and so progress much faster. Reformers tend to undervalue lectures for mature students. I am able to assign the entire content of BPS (excepting only Sections 4.2 and 4.4) in a course for graduate students in education, and to introduce SAS as well. These students, while not quantitatively strong, are mature (many are over 30 years of age) and hard-working. The outlines below are intended for more typical undergraduate students.

2.2   Course outlines

The sample outlines for courses using BPS are aids for instructors, not strict rules. You should adapt the pace and extent of your course to your students.

OUTLINE I is a semester course for students with relatively low quantitative skills. I am always tempted to go faster than this outline suggests, and always find that when I do the students don't come with me. In particular, I find that each exam, viewed as an opportunity to solidify learning, uses a week. I spend one class on active review. I often distribute a sample exam in class and ask the students to work on the first problem for about 5 minutes, long enough to determine whether they know how to approach it. We then discuss that problem together. Continue through the sample exam in this manner. Starting work under conditions similar to an exam concentrates the mind. The exam itself occupies a second class period, and returning and discussing it fills most of a third. This isn't lost time: exams are learning tools. I recommend omitting all starred subsections for this audience, but use your judgment.

OUTLINE II is a semester course for somewhat better-prepared students. This is essentially the content of Purdue's statistical methods course for general (non-calculus) students. The largest clients for this course are agriculture and health sciences. We present inference for regression (Chapter 10 of BPS) as the concluding topic; you may prefer to substitute Chapters 8 or 9. In addition to one of the concluding chapters, this course adds Sections 2.5 (categorical data), 4.2 (probability distributions), and 4.6 (control charts) to the material of OUTLINE I. You may prefer to substitute Section 5.4 (Type I and Type II errors), or to omit all of these and assign either Chapter 8 or Chapter 9.

OUTLINE III describes a one-quarter course. It covers the essentials, ending with the t procedures (and the introduction to inference in practice) in Chapter 6 of BPS. This provides a good foundation for any of a number of more specialized courses in statistics or related areas. This outline is quite tentative; the number of class meetings per week varies in institutions using the quarter system, and I have not taught in such institutions. You may be able to assign more than is shown, e.g., Sections 4.3 and 7.1. The main point of this outline is the goal of completing Chapter 6 in even the briefest introduction.

OUTLINE I: 15 Weeks, Lower skills

Week	Assignment
1	Course introduction; BPS Section 1.1
2	BPS Section 1.2
3	BPS Section 1.3
4	BPS Sections 2.1, 2.2
5	BPS Sections 2.3, 2.4
6	Review of data analysis, EXAM I
7	BPS Sections 3.1, 3.2
8	BPS Sections 4.1, 4.3
9	BPS Sections 4.5, 5.1
10	BPS Sections 5.2, 5.3
11	Review of inference ideas, EXAM II
12	BPS Section 6.1
13	BPS Section 6.2, first subsection in Section 6.3
14	BPS Sections 7.1, 7.2
15	Course review and extended examples
	Comprehensive Final Exam

OUTLINE II: 15 Weeks, Moderate skills

Week	Assignment
1	BPS Sections 1.1, 1.2
2	BPS Section 1.3
3	BPS Sections 2.1, 2.2, start 2.3
4	BPS Sections 2.3, 2.4, 2.5
5	BPS Sections 3.1, 3.2
6	Review of Chapters 1 to 3; EXAM I
7	BPS Sections 4.1, 4.2
8	BPS Sections 4.3, 4.5
9	BPS Section 4.6, 5.1
10	BPS Sections 5.2, 5.3
11	Review of Chapters 4 and 5; Exam II
12	BPS Sections 6.1, 6.2
13	BPS Sections 7.1, 7.2
14	BPS Chapter 10
15	Course review and extended examples
	Comprehensive Final Exam

OUTLINE III: 10 Weeks, Low or Moderate skills

Week	Assignment
1	BPS Sections 1.1, 1.2 through mean and median
2	BPS Sections 1.2, 1.3
3	BPS Sections 2.1, 2.2, 2.3 through facts about least squares
4	BPS Sections 2.3, 2.4; Review data analysis
5	Exam on Chapters 1 and 2; BPS Section 3.1
6	BPS Sections 3.2, 4.1
7	BPS Sections 4.5, 5.1
8	BPS Sections 5.2, 5.3
9	BPS Sections 6.1, 6.2
10	Course review or selected topics
Comprehensive Final Exam	

3   CHAPTER COMMENTS

The comments below contain brief discussions of philosophy, teaching suggestions, and additional data and examples for use in teaching.

3.1   Part I : Understanding Data

One of the most noteworthy changes in statistics instruction in the past decade is the renewed focus on helping students learn to work with data. The change in instruction follows a change in research emphases. Statistics research has pulled back a bit from mathematics (though, as the wise saying goes, you can never be too rich or too thin or know too much mathematics) in favor of renewed attention to data analysis and the problems of scientific inference. It is no longer thought proper to devote a week to "descriptive statistics" (means, medians, and histograms) before plunging into probability and probability-based inference. Contemporary introductions to statistics include a substantial dose of "data analysis." In addition to reflecting statisticians' consensus view of the nature of their subject, working with data has clear pedagogical advantages. Students who may be a bit anxious about the study of statistics can begin by learning concrete skills and exercising judgment that amounts to enlightened common sense.

Chapters 1 and 2 present the principles and some of the tools of data analysis. For teachers who are mathematically trained (as I was), effective teaching of data analysis requires some reorientation. Here are three principles.

1. Emphasize the strategy, not just the skills. It is easy to treat data analysis as a longer stretch of descriptive statistics. Now we present stemplots, boxplots, the 5-number summary, ..., in addition to means, medians, and histograms. There is a larger strategy for looking at data which these tools help implement. I have tried to stress some elements of this strategy, such as

- Begin with a graph, move to numerical descriptions of specific aspects of the data, and (sometimes) to a compact mathematical model. *Which* graphs, numerical summaries, and mathematical models are helpful depends on the setting.

- Look for an overall pattern and for striking deviations from that pattern. Deviations such as outliers may influence the choice of descriptive summaries, and the presence and clarity of the overall pattern suggests what mathematical models may be useful.

2. Don't import inferential ideas too soon. The point of view of data analysis is to let the data speak, to examine the peculiarities of the data in hand without at first asking if they represent some wider universe or answer some broader question. The distinction between sample and population, which is central to inference, is deliberately ignored in data analysis. John Tukey of Bell Labs and Princeton, who shaped the subject, refers to "bunches" of data. I haven't gone that far, but I do delay the sample-population distinction until Chapter 3, where it is essential to the discussion of designs for producing data. One aspect of successful teaching is to resist the temptation to tell students everything at once. Let them grasp the strategy and tools of basic data analysis first. These will be helpful ancillaries when we come to inference.

3. Use real data. Remember the mantra: data are not just numbers; they are numbers with a context. The context enables students to communicate conclusions in words and to judge whether their conclusions are sensible. Data come with at least a bit of background, though for beginning instruction that background may not fully reflect the complexities of the real world. I'm willing to oversimplify for the sake of clarity, but not to ask empty operations with mere numbers.

I have tried to provide small and moderate-size data sets in adequate number for basic instruction. You should want more. Here are some suggestions.

- Get D. J. Hand, F. Daly, A. D. Lunn, K. J. McConway, and E. Ostrowski, *A Handbook of Small Data Sets* (Chapman and Hall, London, 1994). This book, by faculty at Britain's Open University, contains "about 500 real small data sets, with brief descriptions and details of their sources." Moreover, an accompanying DOS disk contains the data in ASCII files that any IBM-compatible text editor can read. Many other books contain more specialized data sets. For example, the data disk that accompanies H. Jean Thiébaux, *Statistical Data Analysis for Ocean and Atmospheric Sciences* (Academic Press, San Diego, 1994) has an excellent collection of data on the subject of its title.

- Mine the electronic terrain. Many data sets are available by anonymous ftp or other electronic means of retrieval. Details go out of date quickly, but here are some suggestions. Carnegie-Mellon University maintains *StatLib*, the preeminent electronic repository of things of statistical interest, including data sets. To get started, send the email message

    send index

  to the address

    statlib@lib.stat.cmu.edu

  The *Journal of Statistics Education*, which is an electronic journal not published in hard copy, includes a section on data sets and maintains these in an archive at North Carolina State University. The *JSE* archive contains much else of interest to teachers of statistics. To receive information send the email message

    send index
    send access.methods

  to the address

    archive@jse.stat.ncsu.edu

  Ohio State University, with NSF support, is developing an Electronic Encyclopedia of Examples and Exercises (EEEE) to be available on a disk and eventually online. EEEE provides quite a bit of background for its data sets and includes student exercises. For information, contact Professor William Notz of the Ohio State Statistics Department.

- Amass your own collection of data. Data about the states, with $n = 50$ or $n = 51$, are a convenient size for simple data analyses. The *Statistical Abstract of the United States* is a good place to start. The *Information Please Environmental Almanac*, which is careful to include the provinces of Canada as well as the states, has much data of interest to students. In the 1994 edition, for example, consider the percent of solid waste output that is recycled (Minnesota is an outlier), toxic chemical releases (Louisiana and Texas are outliers), or per capita energy use in Canada (Alberta is an outlier). The "Almanac" issue of the *Chronicle of Higher Education*, published each year around September 1, contains much data on students and education.

3.2   Chapter 1 Examining Distributions

Students taking a first course in statistics often do not know what to expect. Some may view statistics as a field where the major task is to tabulate large collections of numbers accurately. Others have heard that statistics is more like mathematics with a lot of complicated formulas that are difficult to use. Few are expecting a course where they need to use their common sense and to think.

Your presentation of the material in Chapter 1 sets the tone for the entire course. We would like students to see that they can succeed and to become accustomed to making judgments and discussing findings rather than just solving problems. Try to use selected examples or exercises as a basis for class discussion. Presenting new data of special interest to your students is useful. Don't speed through the descriptive material because it seems simple—students don't always find the mechanics simple, and are not accustomed to "reading" graphics. And they are certainly not used to talking about what the data show.

Section 1.1 Displaying distributions with graphs. Be flexible in assessing student graphs: it isn't always clear whether to split stems in a stemplot or how to choose the classes for a histogram. Try by your flexibility to help students not to get hung up on minor details of graphing. Similarly, how symmetric a histogram or stemplot must be to warrant calling the distribution "symmetric" is a matter for judgment. So is singling out outliers. Be flexible, but discourage students from, for example, calling the largest observation an outlier regardless of whether it is isolated from the remaining observations. Flexibility may also help students live with software. In making stemplots, for example, some software packages truncate long numbers and others round; some put the larger stems on top and others put the smaller stems there. These variations have little effect on our picture of the distribution.

Section 1.2 Describing distributions with numbers. The common descriptive measures summarize things we can see graphically, but they summarize only part of what we can see. The graphical presentation is primary for data analysis.

Students should have a calculator that gives them $\bar{x}$ and s from keyed-in data. Do warn them that many calculators offer a choice between dividing by n and dividing by $n-1$ in finding the standard deviation s. We want $n-1$. (What is worse, many calculators label their choices as σ_n and σ_{n-1}. We haven't met σ yet, but we want to use s to denote the standard deviation of a set of data.) Ask students to do Example 1.11 or Exercise 1.29 to check their calculator skills. Because students should use calculators, I give only the "defining formula" for s, the formula that shows (at least to those who can read algebra) what s is. The "computing formula" based on sums of squares is an anachronism, and does not appear.

If you use software, you may find versions of a boxplot and rules for calculating quartiles that differ slightly from those in BPS. Encourage students to ignore this and to work with what the software reports. Do remember that no single numerical summary is appropriate for all sets of data, and that any numerical summary may miss important features such as gaps or multiple peaks. Exercise 1.31 offers an example.

Example. Table 1 gives the weights of the players on a Big Ten football team, along with their positions. (These are the program weights for Purdue's 1994 team.) These data appear on the Data Disk as the file guide1.dat. We can explore how much football players weigh and how weights vary among positions and within a position. Here are Minitab's descriptive statistics for these data:

	POS	N	MEAN	MEDIAN	TRMEAN	STDEV	SEMEAN
WT	WR	10	180.90	180.00	179.63	8.94	2.83
	OB	16	211.56	212.50	211.00	15.74	3.93
	OL	19	271.63	265.00	269.71	24.95	5.72
	DB	12	187.83	184.00	187.90	10.84	3.13
	LB	8	224.88	225.00	224.88	5.03	1.78
	DL	15	252.53	251.00	252.38	20.56	5.31

	POS	MIN	MAX	Q1	Q3
WT	WR	170.00	202.00	175.00	184.25
	OB	190.00	241.00	196.25	223.00
	OL	236.00	340.00	256.00	288.00
	DB	170.00	205.00	180.25	196.25
	LB	215.00	231.00	223.25	229.50
	DL	220.00	287.00	235.00	265.00

The side-by-side boxplots in Figure 1 plot outliers separately using the $1.5 \times IQR$ rule described in Exercise 1.71. They show the expected effects of position and one very heavy offensive lineman, but also at least one surprise: why is there so little variation among the weights of linebackers?

Section 1.3 The normal distributions. Note that normal distributions are introduced here as models for the overall pattern of some sets of data, and not in the context of probability theory. Although this ordering of material is unusual, it has several advantages. The normal distributions appear naturally in the description of large amounts of data, so that the later assumption for inference that "the population has a normal distribution" becomes clearer. Moreover, mastering normal calculations at this point reduces the barrier posed by the material on sampling distributions and probability (Chapter 4). If the students already know how to compute normal "probabilities" and have some understanding of the relative frequency interpretation from this section, the transition to ideas about probability is easier. It is also true

Table 1. Player weights (pounds) and positions for a football team

Weight	Position	Weight	Position	Weight	Position	Weight	Position
175	WR	170	WR	202	WR	175	WR
184	WR	177	WR	185	WR	183	WR
183	WR	175	WR	190	OB	200	OB
215	OB	220	OB	190	OB	205	OB
233	OB	195	OB	220	OB	210	OB
224	OB	241	OB	216	OB	190	OB
210	OB	226	OB	257	OL	256	OL
278	OL	236	OL	340	OL	265	OL
300	OL	267	OL	304	OL	288	OL
265	OL	260	OL	273	OL	260	OL
290	OL	254	OL	241	OL	280	OL
247	OL	181	DB	197	DB	183	DB
183	DB	204	DB	185	DB	178	DB
180	DB	205	DB	194	DB	170	DB
194	DB	231	LB	225	LB	225	LB
225	LB	231	LB	223	LB	215	LB
224	LB	235	DL	265	DL	287	DL
234	DL	256	DL	232	DL	251	DL
256	DL	245	DL	236	DL	220	DL
255	DL	283	DL	286	DL	247	DL

Position key: WR = wide receiver, OB = offensive back, OL = offensive lineman, DB = defensive back, LB = linebacker, DL = defensive lineman.

Figure 1: Weights and positions of football players

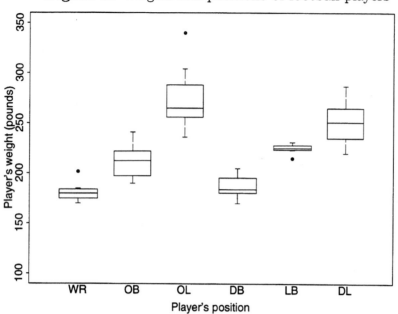

that meeting normal distributions early explains the otherwise mysterious affection of statisticians for the standard deviation.

The organizing idea is that we can sometimes use a mathematical model as an approximation to the overall pattern of data. Normal distributions are one example; a linear regression line (next chapter) is another. The 68–95–99.7 rule is a useful device for interpreting μ and σ for normal distributions. It also makes it possible to think about normal distributions without a table. Many distributions are nonnormal, so don't make this into the so-called "empirical rule" for any distribution.

3.3   Chapter 2 Examining Relationships

Having dealt with methods for describing a single variable, we turn to relationships among several variables. At the elementary level of BPS, that means mostly relationships between two variables. That a relationship between two variables can be strongly affected by other ("lurking") variables is, however, one of the chapter's themes. Note the new vocabulary (explanatory and response variables) in the chapter Introduction, as well as the reiteration of basic strategies for data analysis.

Correlation and regression are traditionally messy subjects based on somewhat opaque "computing formulas" full of sums of squares. BPS asks that students have a "two-variable statistics" calculator that will give them the correlation and the slope and intercept of the least-squares regression line from keyed-in data. This liberates the instructor—we can give reasonably realistic problems and concentrate on intelligent use rather than awful arithmetic. The computing formulas are anachronistic and don't

Table 2. IQ score and grade point average for seventh-grade students

IQ	GPA	IQ	GPA	IQ	GPA	IQ	GPA	IQ	GPA
111	7.940	107	8.292	100	4.643	107	7.470	114	8.882
115	7.585	111	7.650	97	2.412	100	6.000	112	8.883
104	7.470	89	5.528	104	7.167	102	7.571	91	4.700
114	8.167	114	7.822	103	7.598	106	4.000	105	6.231
113	7.643	109	1.760	108	6.419	113	9.648	130	10.700
128	10.580	128	9.429	118	8.000	113	9.585	120	9.571
132	8.998	111	8.333	124	8.175	127	8.000	128	9.333
136	9.500	106	9.167	118	10.140	119	9.999	123	10.760
124	9.763	126	9.410	116	9.167	127	9.348	119	8.167
97	3.647	86	3.408	102	3.936	110	7.167	120	7.647
103	0.530	115	6.173	93	7.295	72	7.295	111	8.938
103	7.882	123	8.353	79	5.062	119	8.175	110	8.235
110	7.588	107	7.647	74	5.237	105	7.825	112	7.333
105	9.167	110	7.996	107	8.714	103	7.833	77	4.885
98	7.998	90	3.820	96	5.936	112	9.000	112	9.500
114	6.057	93	6.057	106	6.938				

appear in the text. Do remember that data input and editing can be frustrating on a calculator, so reserve large problems for computer software.

The descriptive methods in this chapter, like those in Chapter 1, correspond to formal inference procedures presented later in the text. Many texts delay the descriptive treatment of correlation and regression until inference in these settings can also be presented. There are, I think, good reasons not to do this. By carefully describing data first, we emphasize the separate status and greater generality of data analysis. We help students avoid using inference procedures where they clearly do not apply—to data for the 50 states, for example. Fitting a least squares line is a general procedure, while using such a line to give a 95% prediction interval requires additional assumptions that are not always valid. In addition, students become accustomed to examining data *before* proceeding to formal inference, an important principle of good statistical practice. Finally, correlation and regression are so important that they should certainly appear in a first course even if you choose not to discuss formal inference in these settings.

Because elementary "data analysis" for relationships between categorical variables consists mainly of calculating and comparing percents, you can choose to delay the descriptive material of Section 2.5 to accompany the inference methods (the chi-square test) in Chapter 8. If you don't plan to discuss the chi-square test, you can choose to omit Section 2.5 altogether.

Figure 2: IQ and grade point average for seventh grade students

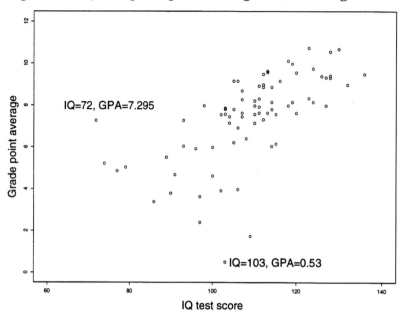

Section 2.1 Scatterplots. Using graphs should be comfortable by now. Constructing scatterplots is a relatively easy task. (But tedious for all but small data sets without software.) Interpreting them, on the other hand, is an art that takes some practice. In the classroom, build instruction on examples and stress that common sense and some understanding of the data are necessary to do a good job of description. Computers can make the plots, but people are needed to describe them. Again, the general rule is to look for overall patterns and deviations from them. Patterns such as clusters and positive and negative association are useful in many cases but can lead to distorted descriptions when imposed in situations where they do not apply.

Example. Table 2 gives data that can be used to illustrate scatterplots, correlation, and regression. The table records the IQ test score and grade point average of 78 seventh grade students from a rural midwestern school. These data make up the file guide2.dat on the data disk.

You may first be interested in the distributions of the two variables. Here's S-PLUS output that describes them:

```
IQ:
Mean    =    108.9
Standard deviation   =    13.17
N = 78   Median = 110
Quartiles = 103, 118


Decimal point is 1 place to the right of the colon
```

```
Low:    72

   7 : 4
   7 : 79
   8 :
   8 : 69
   9 : 0133
   9 : 6778
  10 : 0022333344
  10 : 555666777789
  11 : 0000111122223334444
  11 : 55688999
  12 : 003344
  12 : 677888
  13 : 02
  13 : 6
```

```
GPA:
Mean    =    7.447
Standard deviation   =    2.1
N = 78   Median = 7.829
Quartiles = 6.231, 8.998
```

```
Decimal point is at the colon
Low:   0.53

   1 : 8
   2 : 4
   3 : 4689
   4 : 0679
   5 : 1259
   6 : 0112249
   7 : 22333556666666688899
   8 : 0000222223347999
   9 : 002223344556668
  10 : 01678
```

S-PLUS (using the $1.5 \times IQR$ criterion discussed in Exercise 1.71) identifies one suspected low outlier in each of the variables. These are printed above the stemplots, marked "Low." Notice that IQ (score on a standardized test) has a more symmetric distribution than does GPA, which is concentrated at the higher grades. A normal probability plot (not discussed in BPS) suggests that the distribution of IQ is quite close to normal despite the outlying low scores.

Figure 2 shows the scatterplot of GPA versus IQ. There is a clear linear pattern with moderate positive association. Two outlying points are
marked, one with a very low IQ but good grades and one with average IQ
and extremely poor grades. A third point (IQ 109, GPA 1.76) also falls
in the latter category. We should check that these points are correctly en-
tered; the graduate student in education who collected these data believes
that they are correct. So we have no reason to discard these points.

Section 2.2 Correlation. I present correlation before regression in part because
it does not require the explanatory-response distinction. But my real reason is that
I want to give a meaningful formula for the regression slope, and that requires the
correlation. Students should have a calculator that gives r from keyed-in data. You
can therefore use the somewhat messy formula as a basis for explaining how r behaves
(fit this to your students' ability to read algebra), but avoid using it for substantial
computations.

Example. Return to the data on IQ and GPA in Table 2. The correlation
is $r = .6337$ for all 78 students, $r = .7398$ if we omit all three suspicious
points. It should be clear to students why omitting points that fall far from
the linear pattern *increases* the correlation. Linear association with IQ
predicts about 40% of the observed variation in grades for all 78 students,
about 55% if we drop the three possible outliers.

Section 2.3 Least-squares regression. The background to regression isn't always
clear to students, so don't skip over it: we'd like to draw the *best* line through the
points on our scatterplot; to do this, we need an explicit statement of what we mean
by "best;" the least-squares idea gives such a statement, one that assumes we want
to use the line to predict y from x. Least-squares isn't terribly natural. At this
point, just say that it's the most common way to fit a line. (Least-squares is easily
influenced by extreme observations, but it has many nice properties that have kept
it the standard method even though computers have reduced its ease-of-computation
advantage.)

Using a calculator for the arithmetic allows us to skip the usual rather opaque formulas
for the least-squares line in favor of the wonderful expression

$$b = r \frac{s_y}{s_x}$$

for the slope. This formula tells the algebra-literate a lot, but assess your students
before expounding on it. The concepts of "outlier" and "influential observation" are
important. An observation is influential if removing it would move the regression
line. This is clearly a matter of degree. More advanced statistical methods include
numerical measures of influence. I've defined "outlier" broadly to keep things simple
for students; they only have to look for isolated extreme points in any direction.
That's a matter of degree also.

Example. The least-squares line for predicting GPA from IQ for the data in Table 2 is

$$\hat{y} = -3.5577 + .10103x$$

if we include all the students and

$$\hat{y} = -4.2738 + .10860x$$

when the suspicious observations are discarded. These points have little influence on the regression line because they are not outliers in x. By the way, the least-squares regression line of IQ on GPA is a very different line,

$$\hat{IQ} = 79.326 + 3.9743\text{GPA}$$

That's another reminder that regression requires that you choose the explanatory variable.

Here are the residuals from the least-squares regression of GPA on IQ for the data in Table 2, rounded to three decimal places. The observation numbers at the left identify the first observation in each row; these numbers go down successive columns in Table 2. S-PLUS prints the residuals across rows, so the layout doesn't match Table 2.

```
[1]   0.283 -0.476  0.520  0.207 -0.216  1.205 -0.781 -0.683  0.793
[10] -2.596 -6.319  1.033  0.032  2.116  1.654 -1.903  1.039 -0.007
[19]  0.094 -0.138 -5.695  0.054  0.676  2.015  0.237 -1.723 -1.888
[28] -0.516  0.394  0.440 -1.715  0.219 -1.903 -3.831  0.217  0.749
[37] -0.935 -0.364 -0.795  1.776  1.005 -2.812  1.457  0.638  1.318
[46]  1.461 -0.206 -0.214  0.217 -0.546  0.823 -3.152  1.789  1.726
[55] -1.274  1.534  0.074 -0.389  3.578 -0.290  0.774  0.984  1.242
[64]  0.922  1.125 -0.936 -0.820  1.123  1.005 -0.042  1.891 -0.298
[73] -0.919  1.281  0.679 -0.425  0.663  1.742
```

The sum of these residuals is -0.003, not quite zero due to roundoff error. The suspicious observations are numbers 11, 21, and 59 in this list. They are suspected outliers, so they are identified by large residuals.

Example. The Boston Marathon is one of the world's best-known foot races. The winning time in the Boston Marathon has decreased as runners get faster. Table 3 shows the winning times, in minutes, for the years from 1959 to 1980. (The winning time has not improved much in more recent years, so we don't give those data.)

These data appear on the data disk as file guide3.dat. You can use them for a variety of class examples or student work. A scatterplot (Figure 3) shows a rough linear pattern with quite a bit of year-to-year variation.

Table 3. Winning time (minutes) for the Boston Marathon, 1959–1980

Year	Time	Year	Time
1959	143	1970	131
1960	141	1971	139
1961	144	1972	136
1962	144	1973	136
1963	139	1974	134
1964	140	1975	130
1965	137	1976	140
1966	137	1977	135
1967	136	1978	130
1968	142	1979	129
1969	134	1980	132

Figure 3: Boston Marathon winning times, 1959–1980

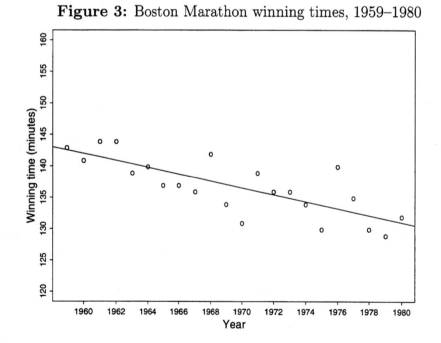

Nonetheless, $r^2 = 0.5993$. Ask students to suggest some factors that help explain the variation about the line. (Weather is one.) The least-squares regression line is

$$\text{Time} = 1221.05 - 0.5505 \times \text{year}$$

This line appears in Figure 3. By how much on the average did the winning time improve per year during this period? Use the line to predict the winning time in 1990, a decade later. Is this prediction trustworthy? (Beware extrapolation. The actual 1990 winning time was 128 minutes.)

Section 2.4 Interpreting correlation and regression. For now at least, computers can't do anything in this section. As calculations are automated, interpretive ideas become a more important part of even basic instruction.

Section 2.5 Relations in categorical data.* This is "applied arithmetic" but students don't find it easy. There is no recipe (I do give guidelines) for deciding what percents to calculate and compare in describing a relationship between two categorical variables. I usually do not assign this section to weaker groups of students. If you will do Chapter 8, you may wish to delay this section and assign both description and inference together.

3.4   Chapter 3 Producing Data

This is a relatively short chapter with a lot of ideas and little numerical work. Students find the essentials quite easy, but they are very important. This chapter isn't mathematics, but it is core content for statistics. Weaknesses in data production account for most erroneous conclusions in statistical studies. The message is that production of good data requires careful planning. Random digits (Table B) are used to select simple random samples and to assign units to treatments in an experiment. There are numerous examples that can serve as the basis for classroom discussion.

The chapter also has a secondary purpose: the use of chance in random sampling and randomized comparative experiments motivates the study of chance behavior in Chapter 4. I have tried to motivate probability by its use in statistics, and to concentrate on the probabilistic ideas most directly associated with basic statistics. This chapter starts that process.

Section 3.1 Designing samples. The deliberate use of chance to select a sample is the central idea. Many of the inference procedures in later chapters assume that the data are a simple random sample. Others require several independent SRSs or another simple model. In this section we learn what an SRS is, and also get a glimpse of the practical difficulties that can damage a sample to the point that formal inference is of little value.

Example. A Roper poll carried out in 1992 for the American Jewish Committee seemed to suggest that many Americans doubted that the Holocaust happened. The exact question asked was:

> *Does it seem possible or does it seem impossible to you that the Nazi extermination of the Jews never happened?*

Of the respondents, 22% said "possible" and 12% said "not sure." After much anguished editorial comment, it was suggested that the convoluted wording of the question had confused some people. So in 1994, Roper conducted a second poll. This time, the question was:

> *Does it seem possible to you that the Nazi extermination of the Jews never happened, or do you feel certain that it happened?*

Now only 1% said it was possible it never happened and 8% said "not sure." When questioned in plain language, Americans know that the Holocaust did occur.

Section 3.2 Designing experiments. I call the randomized comparative experiment the single greatest contribution of statistics to the advance of knowledge. Since Fisher introduced randomization in the 1920s, these ideas have revolutionized the conduct of studies in fields from agriculture to medicine. No student should leave a first statistics course without understanding the distinction between experiments and observational studies and understanding why properly designed experiments are the gold standard for evidence of causation. When experiments can't be done, causation is a slippery subject, and statistical methods that at least claim to give evidence for causation are not for beginners and are often debated even by experts. Good experiments allow relatively clean conclusions.

Example. Here's an example of a randomized comparative experiment to evaluate social policy. The subjects were 300 people convicted three times in one year of drunken driving in San Diego, California. All were given a fine, a suspended jail sentence, and assigned by the judge to one of three treatments: no additional treatment, attending an alcoholism clinic, or attending Alcoholics Anonymous. Here is the design:

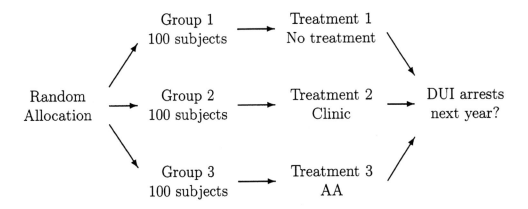

The results were a bit discouraging: 44% of the control group had no DUI arrests in the next year, compared with 32% of both the clinic group and the AA group. These percentages do not differ significantly, that is, they are within the margin of expected chance variation. But the experiment certainly did not show that either treatment reduced future drunk driving.

The San Diego experiment also illustrates the difficulty of doing experiments on social policy: there were no followup data on 20% of the subjects, who left the area or otherwise disappeared during the year.

Example. The *New York Times* (September 30, 1993) carried a report of a study appearing that day in the *New England Journal of Medicine*. The study was based on a large probability sample, a prospective study in which the subjects were followed forward over time. The researchers compared obese women (the top 5 percent on an index relating weight to height) with other women. The study findings raise but don't conclusively settle issues of causation. Here are excerpts from the *Times* article.

Women who are fat suffer enormous social and economic consequences, a new study has shown. They are much less likely to marry than women of normal weight and are more likely to be poor and to earn far less. ...

The findings are from an eight-year study of 10,039 randomly selected people who were 16 to 24 years old when the research began. ...

Fat women were disproportionately found in lower socioeconomic classes, and some researchers say this is because poor women are more likely to eat fat-laden food and junk foods and to get less exercise than richer women. But Dr. Gortmaker and his colleagues wrote, "our data suggest that at least some of this relation may be a socioeconomic consequence of being overweight." ...

The fat women were 20 percent less likely to marry, had household incomes that were an average of $6,710 lower and were 10 percent more likely to be living in poverty.

3.5   Part II : Understanding Inference

The reasoning of classical statistical inference is built on asking, "What would happen if I used this method many times?" Confidence limits, P-values, and error probabilities answer that question in varied settings. All of these answers utilize the *sampling distribution* of a statistic, which addresses the underlying question by displaying the distribution of the statistic in repeated samples or experiments carried out under the same circumstances. Sampling distributions are a tough idea to convey to students, but they are central to inference and can't be avoided without loss of conceptual mastery.

General probability, on the other hand, is a tough subject that we can largely avoid. In basic statistics, our use of probability laws like the "complements rule" is limited to noting that the area in the upper tail of a density curve is one minus the remaining area. *Distributions* are the big idea of probability for understanding the reasoning of basic statistical inference. Distributions are therefore the focus of Chapter 4, where we meet probability. The unstarred sections of Chapter 4 present much important probability content, even the law of large numbers and the central limit theorem, but that content resolutely concentrates on distributions and avoids topics not required for the statistics we will do in later chapters. The optional Sections 4.2 and 4.4 contain additional material on probability, which is not needed to read the remainder of the text, but even this remains focused on probability distributions.

Chapter 5 is no doubt the most difficult in the book, displacing the traditional lump of general probability, which is absent from BPS, from that title by default. There is no hiding the fact that the reasoning of confidence intervals and (more so) significance tests isn't easy. But if all the calculations are done by software, as is now the case in practical applications of statistics, students must carry away this reasoning if our presentation of inference is to have much lasting value.

Students have a limited capacity for hard topics in a course. I recommend applying that capacity to sampling distributions and the reasoning of inference rather than using much of it on general probability. Unless your students are quite well prepared, I suggest skipping the optional sections 4.2, 4.4, and 5.4. In effect, the traditional first course is being modified to place more emphasis on data (Chapters 1, 2, and 3) and concepts (Sections 5.1, 5.2, and 5.3). This is exactly the direction recommended by the ASA/MAA joint curriculum committee. Try it. You may like it.

Chapters 6 and 7 present the simplest inference procedures of interest in practice, for inference about means (Chapter 6) and proportions (Chapter 7). Chapter 6 is essential, because in it we meet many issues relevant to applying statistical methods to real problems. Chapter 7 is not essential. In fact, you can shorten your path through this part of BPS by omitting Section 4.3 and Chapter 7 if you wish. Chapters 9 and 10, but not Chapter 8, are accessible by this route.

3.6   Chapter 4 Sampling Distributions and Probability

Section 4.1 Sampling distributions. Sampling distributions both motivate the study of probability and are the aspect of probability most needed to understand inference.

If you have the capability to do simulations, use it here. Most statistical software packages and many graphing calculators will, for example, simulate the binomial distributions or produce random 0/1 outcomes with stated probability of a 1. That will enable you to have students actually do simulations of SRSs of sizes 100 and 2500 from a population with $p = .6$, at least with enough trials to note sampling variability. Such an exercise will put flesh on the "thought simulations" used in the exposition of this section. They also help make clear the distinction between parameter and statistic.

Example. Here's a simulation activity. Prepare a population of 100 identical small pieces of stiff paper. Write numbers on the slips as follows:

Write each of these numbers	on this many slips
50	10
49, 51	9
48, 52	9
47, 53	8
46, 54	6
45, 55	5
44, 56	3
43, 57	2
42, 58	1
41, 59	1
40, 60	1

You can of course program this simulation on a calculator or computer. Before doing so, ask yourself whether your students really understand simulation. There are advantages to starting with actual physical simulations before moving to the computer, which beginning students often regard as a "magic box."

The 100 slips form a population. The distribution of measurements (the numbers on the slips) in this population is roughly normal with mean $\mu = 50$ and standard deviation $\sigma = 4$. Make a histogram or stemplot of the 100 population values, and find their mean (it is exactly $\mu = 50$ because of the symmetry.) This is a population distribution, and its mean is a parameter.

Now put the slips in a box and have a student take a random sample of size 9 by drawing 9 slips blindly and recording the numbers on them.

Calculate the mean $\bar{x}$ of the 9 observations. This is a statistic. Return these slips to the box and shuffle the slips in the box thoroughly. Draw another random sample of size 9, record the numbers, and find $\bar{x}$. Repeat this as many times as is convenient, preferably about 100 times. Make a histogram of the $\bar{x}$ values, and find their mean and standard deviation. This is an approximation to the sampling distribution of $\bar{x}$.

Probability appears at the end of the section as a language used for shorthand descriptions of what we would observe in an indefinite sequence of trials of a random phenomenon. I say several times how nice it is that mathematics often allows us to bypass actual trials (or simulations of them), and that we will make constant use of facts about probability obtained by mathematical study. But in this setting we avoid the math itself. (At the other end of the sophistication scale, simulation and bootstrapping allow statistics to go beyond the reach of mathematical derivations, so it's perhaps in the spirit of the subject to hide the math from beginners.)

Section 4.2 Probability distributions.* This section expands the brief presentation of probability in the previous section, again emphasizing distributions rather than general probability. It is (by comparison with many aspects of probability) quite straightforward and is helpful in making more exact the meaning of the mean and standard deviation of a distribution. Somehow, however, I always come away feeling that giving the definitions salved my conscience more than it helped my students' understanding. So I have become more willing to omit this material. It isn't required later in the text.

Yes, I know that discrete random variables can take infinitely many values. That's not very helpful to students without a math background that includes infinite series. I like the saying of the physicist Richard Feynman that "The real problem in speech is not precise language. The problem is clear language." He was talking about mathematics textbooks when he said that. We need not tell students everything we know.

Section 4.3 Sample proportions. In this section, we make precise the qualitative facts about the behavior of the sample proportion $\hat{p}$ that were "discovered" in Section 4.1. Despite the natural continuation, the sampling behavior of $\hat{p}$ is a bit messier than that of $\bar{x}$. I tried to write Section 4.5 so that you can go from 4.1 directly to 4.5. I suggest doing that if your course is short or if you want to go more slowly. You must then omit Chapter 7. The more common path includes this section and Chapter 7.

Section 4.4 The binomial distributions.* The inference methods for proportions introduced in beginning statistics (including Chapter 7 of BPS) utilize the normal approximation presented in the previous section. The binomial distributions are therefore an unnecessary complication when our goal is to help students learn basic statistical ideas and methods. I recommend not covering this section (even if you assigned Section 4.2) unless your students' needs or your course description force the binomial upon you.

All the basic facts about the binomial distributions are present, with emphasis on recognizing the binomial setting. The "derivation" of the recipe for binomial probabilities will need parsing for the students. (Or perhaps you will wish to simply omit the derivation.) In particular, the multiplication rule for independent events appears very informally in the derivation. The rule without the informal understanding is just mechanics, and the time needed for a full treatment is better spent elsewhere, so I don't feel too apologetic about this approach. And the rule *is* common sense in specific settings.

Section 4.5 Sample means. This rather short but important section explains the mean and standard deviation of the sample mean and presents the central limit theorem and the law of large numbers. In terms of statistical applications, the two key ideas are stated at the beginning:

- Averages are less variable than individual observations.

- Averages are more normal than individual observations.

I suggest returning to these two ideas at the end of the section, noting how we have made them more precise. This section, along with Section 4.3, completes the discussion of normal distributions begun in Chapter 1 by showing the normal distributions as sampling distributions.

> **Example.** The simulation suggested for Section 4.1 illustrates the fact that $\bar{x}$ is less variable than individual observations. You can demonstrate the central limit theorem effect by a similar simulation. Make (or program the equivalent of) another set of 100 slips, 10 for each number between 1 and 10. This is a uniform distribution. Even though $n = 9$ is not a large sample size, the histogram of $\bar{x}$'s from many random samples of size 9 will look roughly normal.

Section 4.6 Control charts.* Control charts combine graphs and sampling distributions in a simple setting to draw conclusions about a process. They offer an immediate application of facts about sampling distributions without the complications presented by confidence intervals and significance tests. Control charts are therefore an excellent bridge to inference. Nonetheless, they will seem relevant only to some groups of students (particularly business and technical majors) and aren't needed to continue in the text. The exposition tries to explain in simple terms the thinking behind SPC in addition to the workings of the simplest $\bar{x}$ chart.

3.7   Chapter 5 Introduction to Inference

This chapter contains many fundamental ideas. We introduce confidence intervals and tests along with some cautions concerning the use and abuse of tests. Throughout, the setting is inference about the mean μ of a normal population with known standard deviation σ. As a consequence, the z procedures presented are not applicable to most real sets of data. They introduce ideas in a setting where students can do familiar normal calculations, and they pave the way for the more useful t procedures presented in the next chapter.

Experience shows that many students will not master this material upon seeing it for the first time. Fortunately, they will meet the key ideas again in the next chapter. By the time they have completed both chapters and worked many exercises, they should grasp the fundamentals. Be patient, and remember that understanding the reasoning of inference is more important that the number of procedures learned.

Section 5.1 Estimating with confidence Figure 5.3 in BPS displays the big idea: the recipe for a 95% confidence interval produces intervals that hit the true parameter in 95% of all possible samples. (In formal language, the recipe has probability 0.95 of producing an interval that catches the true parameter.) I recommend using simulation to help students understand this central idea.

> **Example.** You can use the same population prepared for the simulation activity of Section 4.1. For random samples of size $n = 9$, the 95% confidence interval for the mean μ is
>
> $$\overline{x} \pm 1.96\frac{\sigma}{\sqrt{n}} \;=\; \overline{x} \pm \frac{4}{\sqrt{9}}$$
> $$\doteq\; \overline{x} \pm 1.33$$
>
> Once again, it is a good idea to begin with several samples drawn by hand. You can program follow-up computer simulation to generate $n = 9$ observations from the normal distribution with mean $\mu = 50$ and standard deviation $\sigma = 4$, find the sample mean $\overline{x}$ of these observations, and calculate $\overline{x} \pm 1.33$. Do this at least 100 times, and observe how many of the intervals cover 50. A drawing like Figure 5.3 of BPS, constructed sample by sample, is very instructive. Note that it is unlikely that exactly 95 of 100 samples cover the true μ, even though 95% will cover in the long run.

Section 5.2 Tests of significance The reasoning of significance tests is conceptually the hardest point in a first course in statistics. Figure 5.8 of BPS is my best attempt to convey the core idea: one $\overline{x}$ value in that figure isn't surprising if the true mean μ is 0, and the other is. If we got the surprising value, we would doubt that μ really is 0. Stripped of all jargon, this reasoning is fairly straightforward *if* the distinction between parameter and statistic is firmly in place. Students do need to know the standard terminology and organization of significance tests, which can nearly obscure the core idea.

> **Example.** The diastolic blood pressure for American women aged 18 to 44 has approximately the normal distribution with mean $\mu = 75$ millimeters of mercury (mm Hg) and standard deviation $\sigma = 10$ mm Hg. We suspect that regular exercise will lower the blood pressure. A sample of 25 women who jog at least 5 miles a week gives sample mean blood pressure $\overline{x} = 71$ mm Hg. Is this good evidence that the mean blood diastolic blood pressure for the population of regular exercisers is lower than 75 mm Hg?
>
> The alternative is one-sided because we suspect that exercisers have *lower* blood pressure.
>
> $$H_0 : \mu = 75$$
>
> $$H_a : \mu < 75$$
>
> Assuming that joggers have the same σ as the general population, the z statistic is
>
> $$z = \frac{\overline{x} - \mu_0}{\sigma/\sqrt{n}}$$
>
> $$= \frac{71 - 75}{10/\sqrt{25}} = -2.00$$
>
> Sketch a standard normal curve. How surprising is a z this small? The 68–95–99.7 rule says it's quite surprising. More formally, the P-value is the probability that z takes a value this small or smaller. This is $P(Z \leq -2.00)$. From Table A, $P = 0.0228$. This result *is* significant at the 5% level ($\alpha = .05$), but is *not* significant at the 1% level ($\alpha = .01$).

Section 5.3 Using significance tests In discussing z confidence intervals, I offer a "warning label" reminding users of conditions for proper use. That label applies to the z tests also. Tests are, however, more difficult to interpret than are confidence intervals. Many statisticians feel that tests are overused, or at least over-interpreted. Hence this short section. The discussions titled "Choosing a level of significance" and "Statistical significance and practical significance" offer some cautions about the interpretation of statistical significance. "Statistical inference is not valid for all sets of data" and "Beware of multiple analyses" apply to confidence intervals as well, but abuses seem more common in the setting of tests.

Section 5.4 Inference as decision* Some instructors stress P-values in teaching beginners; others stress the two types of error and their associated error probabilities. I'm in the former camp, though I didn't start out there. Why begin with P-values? First, I think "assessing the strength of evidence" is a better description of practical inference than is "making decisions." Second, P-values are prominent in the output from statistical software, so users of statistics must understand them. The fact that there is an elegant mathematical theory (Neyman-Pearson) based on the fixed-α approach should not be allowed to sway practical instruction for beginners.

That said, P-values are not sufficient for a full account of statistical tests. The idea of *power* (how likely is this test to detect an alternative you really want to detect if it is true) is certainly important in practice. To compute a power, we start with a fixed α even if in practice we plan to report a P-value. So teachers must ask: given the actual state of our students, how far should we go on from P-values? I confess that I rarely cover this section with undergraduates.

Your choice may differ. Be aware that the ideas are sophisticated and that the proper approach to inference is still hotly debated. Some of my most able colleagues, for example, deny that P-values "assess the strength of evidence against H_0" at all. They do this because they are Bayesians who (I think) import elements that are not in fact present in a majority of real statistical problems and judge P-values based on these imports. I have tried to organize the material in the order of its importance in current statistical practice. That's why P-values are up front and the two types of error are in an optional section.

3.8   Chapter 6 Inference for Distributions

The one- and two-sample t procedures are among the most-used methods of inference. One-sample t confidence intervals and significance tests are a short step from the z procedures of Chapter 5. The two-sample procedures present a complication: the "textbook standard" method assumes equal variances in the population, an assumption that is hard to verify and often not justified. BPS ignores those methods in favor of two alternatives that work even if the population variances differ: a reasonably good conservative approximation for hand use and a very accurate approximation that is implemented in almost all statistical software packages. You can find justifications for this choice in the literature cited in Note 14 of this chapter in BPS. Another deviation from the "textbook standard" occurs in the optional Section 6.3, where the basic recommendation concerning inference about population spread is "Don't do it without expert advice." This choice is also well justified by literature citations.

The exposition in this chapter pays at least some attention to the problems of applying statistical inference to real data. See for example the discussions following Example 6.2, Example 6.3, Example 6.7, and 6.9, and the sections on the "robustness" of both one- and two-sample t.

Section 6.1 Inference for the mean of a population If you want to do inference about μ but don't know σ, just replace the unknown σ by its sample estimate s in the z procedures. That's the driving idea. It leads to the t distributions and to use of all of Table C. Because the mechanics are so similar to those of Chapter 5, you can replay the reasoning of inference while paying more attention to interpreting the results. The section calls attention to the use of one-sample methods for matched pairs data and to the conditions needed to use the methods in practice.

Example. A milk processor monitors the number of bacteria per milliliter in raw milk received for processing. A random sample of 10 one-milliliter specimens from milk supplied by one producer gives the following data:

5370, 4890, 5100, 4500, 5260, 5150, 4900, 4760, 4700, 4870

Entering these data into a calculator gives $\bar{x} = 4950$ and $s = 268.45$. So a 90% confidence interval for the mean bacteria count per milliliter in this producer's milk is

$$\bar{x} \pm t^* \frac{s}{\sqrt{n}} \;=\; 4950 \pm 1.833 \frac{268.45}{\sqrt{10}}$$
$$=\; 4950 \pm 155.6$$

This interval uses the critical value from the $t(9)$ distribution for $C = 90\%$, found in Table C.

Now ask your students what assumptions this calculation requires and how they would verify the assumptions. We must be confident that the data are an SRS from the producer's milk and that the distribution of bacteria counts in the population of milk is approximately normally distributed. We must learn how the sample was chosen to see if it can be regarded as an SRS. The data show no outliers and no strong skewness; it's hard to assess normality more closely from only 10 observations. In practice we would probably rely on the fact that past measurements of this type have been roughly normal.

Example. The amount of wax deposited on the outside surface of waxed paper bags during production may vary from the amount deposited on the inside surface. Select a sample of 25 bags, determine the wax concentration in pounds per square foot on the inside and outside, and calculate the difference (outside minus inside) for each bag. The mean and standard deviation of these 25 differences are

$$\bar{x} = 0.093 \quad s = 0.723$$

Is there good evidence that the mean concentrations on the two surfaces are not equal?

This is a matched pairs situation. The inner and outer surface of the same bag are paired with each other. We want to test the hypothesis of "no difference," or

$$H_0 : \mu = 0$$

$$H_a : \mu \neq 0$$

where μ is the mean of the difference in wax concentrations on the two surfaces. The one-sample t statistic is

$$t = \frac{\overline{x} - \mu_0}{s/\sqrt{n}}$$

$$= \frac{0.093 - 0}{0.723/\sqrt{25}} = 0.643$$

Compare this value with critical points of the $t(24)$ distribution. Table C shows that $t = 0.643$ is not as extreme as the $p = 0.25$ critical value, which is $t^* = 0.685$. Because the test is two-sided, we double the p from the table to get the P-value. So P is greater than 0.5. There is no evidence that the wax concentrations on the two surfaces differ.

Section 6.2 Comparing two means

Students now need to distinguish one-sample, matched pairs, and two-sample settings. That's how this section of BPS opens. For inference about the difference $\mu_1 - \mu_2$ of two population means, we start with the natural sample estimator $\overline{x}_1 - \overline{x}_2$ and its sampling distribution. The distribution is (at least approximately) normal, so standardize the estimator and replace the unknown σ_i by the sample standard deviations s_i. That's the two-sample t statistic. You may not wish to emphasize this intuitive "derivation," depending on your students' capacities for generalization, but it repeats the logic of earlier settings. We then come to the actual two-sample t procedures: just use the smaller of $n_1 - 1$ and $n_2 - 1$ as the degrees of freedom. Everything else is optional, but if you are using software you will want students to read the section headed "more accurate levels."

Example. In an experiment to study the effect of the spectrum of the ambient light on the growth of plants, researchers assigned tobacco seedlings at random to two groups of 8 plants each. The plants were grown in a greenhouse under identical conditions except for lighting. The experimental group was grown under blue light, the control group under natural light. Table 4 gives the data on stem growth in millimeters. These data form the file guide4.dat on the data disk.

This is a two-sample situation. Call plants grown under natural light Population 1 and plants grown under blue light Population 2. We will give a 90% confidence interval for the amount by which blue light reduces stem growth during this period. A calculator gives

Population	Sample size	Sample mean	Sample variance
1	$n_1 = 8$	$\overline{x}_1 = 4.0875$	$s_1^2 = 0.0270$
2	$n_2 = 8$	$\overline{x}_2 = 3.0125$	$s_2^2 = 0.0298$

Table 4. Stem growth (millimeters) of seedlings under normal and blue light

Control Group		Experimental	
4.3	4.2	3.1	2.9
3.9	4.1	3.2	3.2
4.1	4.2	2.7	2.9
3.8	4.1	3.0	3.1

The 90% confidence interval for the difference $\mu_1 - \mu_2$ between the population means is

$$(\bar{x}_1 - \bar{x}_2) \pm t^* \sqrt{\frac{s_1^2}{n_1} + \frac{s_2^2}{n_2}}$$

$$= (4.0875 - 3.0125) \pm 1.895 \sqrt{\frac{.0270}{8} + \frac{.0298}{8}}$$

$$= 1.075 \pm 0.160$$

$$= (0.915, \ 1.235)$$

The degrees of freedom are 7, because both $n_1 - 1$ and $n_2 - 1$ are 7. The confidence interval uses the $C = 90\%$ critical value of the $t(7)$ distribution, which Table C gives as $t^* = 1.895$.

Example. A study of the effect of eating sweetened cereals on tooth decay in children compared 73 children (Group 1) who ate such cereals regularly with 302 children (Group 2) who did not. After three years the number of new cavities was measured for each child. The summary statistics are:

Group	n	$\bar{x}$	s
1	73	3.41	3.62
2	302	2.20	2.67

The researchers suspected that sweetened cereals increase the mean number of cavities. Do the data support this suspicion?

This is another two-sample situation. We wish to test the hypotheses

$$H_0 : \mu_1 = \mu_2$$
$$H_a : \mu_1 > \mu_2$$

The two-sample t test statistic is

$$t = \frac{\bar{x}_1 - \bar{x}_2}{\sqrt{\frac{s_1^2}{n_1} + \frac{s_2^2}{n_2}}}$$

$$= \frac{3.41 - 2.20}{\sqrt{\frac{3.62^2}{73} + \frac{2.67^2}{302}}} = 2.68$$

To assess the significance of the one-sided test, compare $t = 2.68$ with the upper critical values of the $t(72)$ distribution. The degrees of freedom are the smaller of $n_1 - 1 = 72$ and $n_2 - 1 = 301$. Using the $df = 60$ line in Table C, we see that $P < 0.005$. In particular, the result is significant at the 1% level. There is strong evidence that the sweetened cereal group has more cavities on the average.

Normality of the cavity counts isn't important to the validity of the test, because of the large samples. We do need to know how the samples were chosen. We can't conclude from this study that eating sweetened cereal *causes* more cavities. The children who eat such cereal may differ in many ways from those who don't.

Section 6.3 Inference for population spread* The contrast in the practical usefulness of the t procedures for means and the chi-square and F procedures for standard deviations is a good argument for not allowing theoretical statistics to set the agenda for a first course in statistical methods. These tests are all (at least approximately) likelihood ratio tests for normal distributions. They therefore share a widely accepted general principle and some large-sample optimality properties. But they are vastly different in their actual usefulness. The t tests (and their extension to the ANOVA test for comparing many means) are relatively little affected by deviations from normality. Tests for standard deviations, on the other hand, are so sensitive to deviations from normality that I do not believe they should be used in practice. Footnotes 6 and 29 in BPS point to some of the large literature. Start your reading with the paper of Pearson and Please cited in Note 6. Their graphs effectively display the contrast between tests on means and tests on standard deviations and add several fine points to the brief discussion in BPS. (One of their graphs is reproduced in Section 7.3 of IPS, where the discussion of this issue is naturally more extended than in BPS.)

Should we present the standard tests for standard deviations in a first statistics course? BPS allows three choices. You can ignore these tests altogether (this section is optional). You can discuss the robustness issue and also present the most common of the questionable procedures, the F test for comparing two standard deviations. Or you can discuss only the first subsection in Section 6.3, bluntly titled "Avoid inference about standard deviations," and omit the actual F test on the grounds that we have explained why it is not of much value. I usually take the third approach.

3.9   Chapter 7 Inference for Proportions

This chapter presents the z procedures for one-sample and two-sample inference about population proportions. The procedures are approximate, based on the large-sample normal approximation discussed in Section 4.3. By now the students should be comfortable with the general framework for confidence intervals and significance tests. Those who have not yet mastered these concepts get an additional opportunity to learn these important ideas.

Section 7.1 Inference for a population proportion

Here are confidence intervals and significance tests for a single proportion. There is a slight additional complication concerning the standard deviation of $\hat{p}$. For confidence intervals we use the standard error $\sqrt{\hat{p}(1-\hat{p})/n}$, whereas for tests of $H_0 : p = p_0$ we use $\sqrt{p_0(1-p_0)/n}$. Students will just follow the recipes given, but you may want to point out why this distinction is reasonable. We want to use all of the information available in a problem for inference. For the confidence interval, p is unknown and the standard error must therefore be estimated using the value of $\hat{p}$ obtained from the data. On the other hand, when testing $H_0 : p = p_0$, our calculations are based on the supposition that H_0 is true and we therefore use the value p_0 in the calculations. Note that these choices destroy the exact correspondence between confidence intervals and two-sided tests (reject if the hypothesized parameter value is outside of the confidence interval).

Section 7.2 Comparing two proportions

This section presents confidence intervals and significance tests for comparing two population proportions. Students should be able to distinguish two-sample from one-sample settings from their work in Chapter 6. As in the previous section, we use different standard errors for confidence intervals and tests. Pooling the two samples in the test statistic, while making the test inconsistent with the confidence interval, keeps the two-sample test consistent with the 2×2 case of the chi-square test for two-way tables in Chapter 8.

There are "exact" inference procedures for proportions based on the binomial distributions. The discreteness of the binomial makes these procedures a bit awkward. You can find details in Chapter 2 of Myles Hollander and Douglas Wolfe, *Nonparametric Statistical Methods*, Wiley, New York, 1973.

3.10   Part III : Topics in Inference

The concluding three chapters of BPS present independent accounts of inference in three more advanced settings: two-way tables of count data (Chapter 8), one-way analysis of variance (Chapter 9), and simple linear regression (Chapter 10).

Each of these settings introduces a distinctive new idea. The issue of "multiple comparisons" arises both in Chapter 8 (comparing many proportions) and in Chapter 9 (comparing many means). In fact, the sections headed "The problem of multiple comparisons" are identical in both chapters except that one mentions proportions and the other means. Chapter 10 must face the fact that the data are no longer one or several simple random samples. The section headed "The regression model" presents the model of random variation about a non-random relationship of known form (a line) but unknown parameters. In all three chapters, however, I have refrained from introducing new types of inference methods such as formal multiple comparisons procedures or confidence bands for the entire true regression line. These chapters are a capstone for a first course and perhaps an introduction to further study. The methods are still the confidence intervals and tests whose logic is the heart of a basic introduction to statistical inference.

3.11   Chapter 8 Inference for Two-Way Tables

The Pearson chi-square test is one of the most common inference procedures and, because it tests the existence of a relationship between two categorical variables under several sampling models, one of the most versatile. Do note the stress that the overall test ("Yes, these variables are related") is not a full analysis of the data. The descriptive analysis of the nature of the relationship is essential.

> **Example.** The type of medical care that a patient receives sometimes varies with the age of the patient. For example, women should receive a mammogram and biopsy of any suspicious lump in the breast. Here are data from a study that asked whether women did receive these diagnostic tests when a lump in the breast was discovered.

	Tests done?	
Age	No	Yes
45–64	61	158
65–74	40	103
75–90	53	77

In this study, a single sample was classified two ways, by age and by whether or not the tests were done. The resulting 3×2 table is similar to the 3×2 table resulting from Example 8.1 in BPS, but the latter example is an experiment in which the number of subjects who received the three treatments was set in advance.

The chi-squared test for this table gives $X^2 = 7.3668$, df $= 2$, $P = 0.0251$. There is quite good evidence that the proportion of women in the population for whom the tests were done differs among the three age groups. We see that the sample proportions are 72% for women aged 45 to 64 years, 72% again for women aged 65 to 74 years, and 59% for women over 75 years. The data do not, of course, show the reason for the observed difference. It may be that doctors judge that surgery or other intervention is too risky for older women in poor health and so do not do the diagnostic tests on such women.

The chi-square test, like the tests in Chapter 7, is an approximate test whose accuracy improves as the cell counts increase. There is an "exact" test for two-way tables, called the Fisher exact test. For example, this test reports $P = 0.0071$ rather than the chi-square test's $P = 0.0052$ for the data of Example 8.1 of BPS. This test treats *both sets of marginal totals* as fixed in advance. In Example 8.1, only one set was fixed by the design of the study. Some statisticians prefer to always do inference "conditional" on the observed marginal totals, as Fisher's test does. This is a debate that you don't want to reveal to your students! You can find a description of the Fisher test in e.g. Alan Agresti, *Categorical Data Analysis* (Wiley, New York, 1990) and a more advanced survey in the same author's "A survey of exact inference for contingency tables," *Statistical Science* 7, 1992, pp. 131–177.

Why does BPS not include the chi-square "test of fit" that assesses e.g. whether the 6 sides of a die really are equiprobable? I don't find this test to be at all common in in practice. Once we escape dice and coins, the test of fit problem usually concerns not fit to a fully specified model but fit to a *family* of models. "Do these data come from a normal population?" is an example. Unfortunately, adapting the chi-square test of fit to such settings is a bit complicated (and is often done incorrectly in texts). In an earlier incarnation I did considerable research on variations of chi-square tests of fit. My survey paper cited in Note 6 in BPS explains the ins and outs of these tests for users. You can find the facts there.

3.12 Chapter 9 Comparing Several Means: One-Way Analysis of Variance

This chapter presents the one-way ANOVA F test, with the descriptive analysis needed to suggest *what differences* among the observed means account for significance of the overall test. Although BPS doesn't do formal multiple comparisons, this chapter (like Chapter 8) points to the problem and avoids suggesting that the ANOVA test is all there is to comparing several means.

You should now assign the part of Section 6.3 that introduces the F distributions if you did not do so earlier.

The algebra of ANOVA is a bit formidable for beginning students—and does not help most students see how the analysis works. I therefore try to explain the nature of the ANOVA test in the section headed "The idea of analysis of variance." This being a more advanced chapter, the algebraic details do appear in the optional section called "Some details of ANOVA." Even here, however, formulas intended for calcu-

Figure 4: Weight gains of rats fed four diets

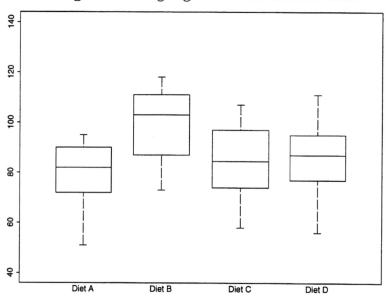

lation with a basic calculator don't appear. I would not make students do ANOVA calculations without software. If you wish to present this chapter to students who lack computer access, I suggest modifying the exercises by giving the sums of squares (see the solutions to the exercises). Students can then find degrees of freedom, mean squares, and the F statistic, and use Table D to assess significance.

Example. In the discussion of Part I, I recommended the *Handbook of Small Data Sets* by D. J. Hand et al. One of their data sets (page 7) records the weight gains (in grams) of 10 rats fed each of four diets. The data come from G. W. Snedecor and G. C. Cochran *Statistical Methods* (6th edition, Iowa State University Press, 1967, page 347), so you can find them in either source. They don't appear here or on the data disk because I didn't ask permission. Figure 4 displays boxplots of the 4 sets of weight gains. The spreads appear similar and there are no outliers (the boxplot would display any suspected outliers using the $1.5 \times IQR$ criterion). Diet B appears to give somewhat higher weight gains than the other three diets.

The basic descriptive statistics are:

	Diet A	Diet B	Diet C	Diet D
n_i	10	10	10	10
$\overline{x}_i$	79.2	100.0	83.9	85.9
s_i	13.89	15.14	15.71	15.02

and output from an ANOVA is

	df	Sum of Sq	Mean Sq	F Value	P value
diet	3	2404.1	801.367	3.58402	0.0229666
error	36	8049.4	223.594		

The degrees of freedom are $I - 1 = 4 - 1 = 3$ in the numerator and $N - I = 40 - 4 = 36$ in the denominator. The P-value $P = 0.023$ indicates good evidence of some differences among the population mean weight gains.

3.13   Chapter 10 Inference for Regression

There are many interesting problems in which the relationship between two variables can be summarized graphically and numerically with a least squares line. Not all of these can be analyzed using the methods presented in this chapter. Inference for linear regression is based on a statistical model that expresses the assumptions underlying the inference procedures. The section headed "The regression model" is therefore essential to understanding regression inference. This section also introduces s, the "standard error about the line," as the key measure of sample variability in the regression setting. You will sometimes find s called "residual standard error" or "root MSE" in computer output or other texts.

The calculations required for regression inference, even after the least-squares line is in hand, are quite unpleasant without software. Most exercises in this chapter therefore give the output from a regression program. If your students are using software, you can ask them to produce the equivalent output from the data. Exercise 10.21 to 10.23 are not feasible without software. If your students lack software access, you can give them the results of key calculations (see the exercise solutions).

Example. A study of the force (y in pounds) require to draw a plow at tractor speed x (miles per hour) gave the following data.

x	0.9	1.3	2.0	2.7	3.4	3.4	4.1	5.2	5.5	6.0
y	425	420	480	495	540	530	590	610	690	680

These data appear on the data disk as the file guide5.dat. Figure 5 shows the scatterplot with the least-squares line. Here is regression output:

```
Residual Standard Error = 19.1869,   Multiple R-Square = 0.9643
N = 10,   F-statistic = 216.0738 on 1 and 8 df,   p-value = 0
```

	coef	std.err	t.stat	p.value
Intercept	362.0657	13.9064	26.0358	0
X	53.3143	3.6270	14.6994	0

Figure 5: Tractor speed (mph) and force required to pull a plow (pounds)

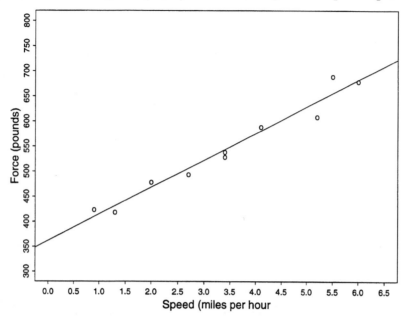

The linear relationship is both very strong ($r^2 = 0.9643$) and highly significant (test for $H_0 : \beta = 0$ has $t = 14.7$ and $P < 0.0001$). The slope β of the true regression line is the rate at which the force required to pull the plow increases with speed. The 95% confidence interval for β is

$$b \pm t^*\mathrm{SE}_b = 53.3143 \pm (2.306)(3.6270) = 53.31 \pm 8.37$$

4   SAMPLE EXAMINATIONS

These sample examinations illustrate typical questions, most of them quite straightforward.

- Sample Examination I covers material from Chapters 1 to 3 of BPS. Most students grasp this content well. Warn them not to become complacent, as later chapters contain more difficult material.

- Examination II covers Chapters 4 and 5. Because these chapters are both important and more difficult than the preceding and following content, I suggest emphasizing them by devoting an exam to them. The sample exam does not contain questions on the optional topics in Sections 4.2, 4.4, and 4.6.

- The third examination is comprehensive, but emphasizes Chapters 6, 7 and 10. (I find that my students can realistically master only one of Chapters 8, 9, and 10.)

You can use these questions as examination items or as class examples. As mentioned earlier, I often use a sample exam distributed in class as review before a scheduled exam.

4.1 Sample examination I

(25) 1. A study examined how long aircraft air conditioning units operated after being repaired. Here are the operating times (in hours) for one unit:

97	51	11	4	141	18	142	68	77
80	1	16	106	206	82	54	31	216
46	111	39	63	18	191	18	163	24

(a) Make a histogram of these data. Take your classes to be 40 hours wide, beginning with

$$0 \le \text{time} < 40$$

$$40 \le \text{time} < 80$$

(b) Describe the overall shape of the distribution: is it roughly symmetric, skewed to the right, or skewed to the left? Are there any outliers?

(c) Is the five-number summary or the mean and standard deviation a better brief summary for this distribution? Explain your choice. Calculate the one of these summaries that you choose.

(20) 2. Biologists and ecologists record the distributions of measurements made on animal species to help study the distribution and evolution of the animals. The African finch *Pyrenestes ostrinus* is interesting because the distribution of its bill size has two peaks even though other body measurements follow normal distributions. For example, a study in Cameroon found that the wing length of male finches varies according to a normal distribution with mean 61.2 mm and standard deviation 1.8 mm.

(a) What proportion of male finches have wings longer than 65 mm?

(b) What is the wing length that only 2% of male finches exceed?

(20) 3. The drug AZT was the first effective treatment for AIDS. An important medical experiment demonstrated that regular doses of AZT delay the onset of symptoms in people in whom the AIDS virus is present. The researchers who carried out this experiment wanted to know

- Does taking either 500 mg of AZT or 1500 mg of AZT per day delay the development of AIDS?

- Is there any difference between the effects of these two doses?

The subjects were 1200 volunteers already infected with the AIDS virus but with no symptoms of AIDS when the study started.

(a) Outline the design of the experiment.

(b) Describe briefly how you would use a table of random digits to do the randomization required by your design. Then use Table B beginning at line 110 to choose *the first five* subjects for one of your groups.

(35) 4. A long-term study of changing environmental conditions in Chesapeake Bay found the following annual average salinity readings in one location in the bay:

Year	1971	1972	1973	1974	1975	1976	1977
Salinity (%)	13.2	9.3	14.9	13.9	14.8	13.3	15.0

Year	1978	1979	1980	1981	1982	1983	1984
Salinity (%)	15.3	15.1	13.1	17.0	19.3	15.6	15.3

(a) Make a plot of salinity against time. Was salinity generally increasing or decreasing over these years? Is there an overall straight line trend over time?

(b) What is the correlation between salinity and year? What percent of the observed variation in salinity is accounted for by straight line change over time?

(c) Find the least-squares regression line for predicting salinity from year. Explain in simple language what the slope of this line tells you about Chesapeake Bay.

(d) If the trend in these past data had continued, what would be the average salinity at this point in the bay in 1988?

4.2   Sample examination I solutions

1. (a) These data appear in the file guide6.dat of the data disk. Figure 6 is the histogram.

   (b) The distribution of operating times is strongly skewed to the right. There are no outliers.

   (c) The five-number summary is preferable for this strongly skewed distribution. First arrange the observations in increasing order:

1	4	11	16	18	18	18	24	31	39	46	51	54	63
68	77	80	82	97	106	111	141	142	163	191	206	216	

   The five-number summary of these $n = 27$ observations is

   $$1 \quad 18 \quad 63 \quad 111 \quad 216$$

Figure 6: Operating hours of aircraft air conditioners

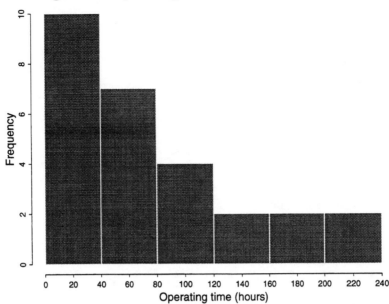

2. (a) Wing length x has the $N(61.2, 1.8)$ distribution. So we want the area under a normal curve such that

$$x > 65$$

$$\frac{x-61.2}{1.8} > \frac{65-61.2}{1.8}$$

$$z > 2.11$$

Table A gives this area as $1 - .9826 = .0174$. About 17.4% of male finches have wing lengths exceeding 64.9 mm.

(b) We want the x with area 0.02 to its right, or area 0.98 to its left. In the body of Table A, find $z = 2.06$ as the entry with left tail area closest to 0.98. So

$$x = 61.2 + (1.8)(2.06) = 64.9 \text{ mm}$$

3. (a) The goals of the experiment require *three* treatment groups, one of which receives a placebo. (Because AZT was the first AIDS drug, it was considered ethical to give a placebo to test its effectiveness. Later drugs were tested against AZT.) Here is the design:

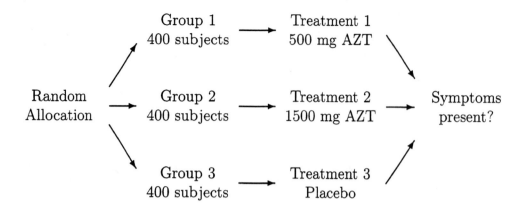

(b) First assign labels. We use labels 0001 to 1200. Then read 4-digit groups from line 110, continuing to the following lines. The first 5 subjects chosen are

$$0676 \quad 0041 \quad 0404 \quad 1197 \quad 0640$$

4. (a) The data are on the data disk as guide7.dat. The plot is Figure 7. There is an increasing linear trend over time.

(b) The correlation (use a calculator) is $r = 0.6386$. Because $r^2 = 0.4079$, linear change over time explains about 41% of the observed variation in salinity over this period.

(c) The least-squares line (use a calculator) is

$$\hat{y} = -659.4385 + 0.340879x$$

That is, salinity is increasing by 0.34% per year on the average. This line is drawn on the scatterplot of Figure 7.

(d) The prediction for $x = 1988$ is

$$\hat{y} = -659.4385 + (0.3409)(1988) = 18.23\%$$

4.3   Sample examination II

(20) 1. About 22% of the residents of California were born outside the United States. You choose an SRS of 1,000 California residents for a sample survey on immigration issues. You want to find the probability that 250 or more of the people in your sample were born outside the U.S.

(a) You would like to use the normal approximation for the sampling distribution of a sample proportion to answer this question. Explain carefully why you can use the approximation in this setting.

(b) What is the probability that 250 or more of the people in your sample were born outside the U.S.?

Figure 7: Chesapeake Bay salinity, 1971–1984 (percent)

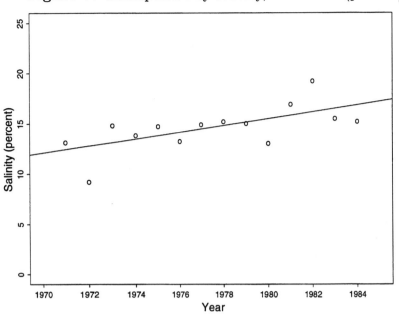

(20) 2. The weights of newborn children in the U.S. vary according to the normal distribution with mean 7.5 pounds and standard deviation 1.25 pounds. The government classifies a newborn as having low birth weight if the weight is less than 5.5 pounds.

    (a) What is the probability that a baby chosen at random weighs less than 5.5 pounds at birth?

    (b) You choose 3 babies at random. What is the probability that their average birth weight is less than 5.5 pounds?

(20) 3. Answer each of the following short questions.

    (a) Give the upper 0.025 critical value for the standard normal distribution.

    (b) An animal scientist is studying factors that affect the level of milk production in dairy cows. He wonders:

        Is the mean production different for cows who are given forage spread on the ground than for cows whose forage is in bunks?

    State the null and alternative hypotheses that you would use in a statistical test of this question. (We can't test these hypotheses yet.)

    (c) An opinion poll asks 1500 randomly chosen U.S. residents their opinion about relations with the Soviet Union. The announced margin of error for 95% confidence is ±3 points. But some people were not on the list from which respondents were chosen, some could not be contacted, and some refused to answer. Does the announced margin of error include errors from these causes?

(d) A student organization plans to ask of 100 randomly selected students whether they favor a change in the grading system.. You argue for a sample of 900 students instead of 100. You know that the standard deviation of the proportion $\hat{p}$ of the sample who say "Yes" will be xxx times as large with the larger sample. Is xxx nine, one-ninth, three, or one-third? Explain your answer.

(e) You read in a journal a report of a study that found a statistically significant result at the 5% significance level. What can you say about the significance of this result at the 1% level: Is it certainly significant, at the 1% level, certainly not significant at the 1% level, or maybe significant and maybe not significant?

(40) 4. A friend who hears that you are taking a statistics course asks for help with a chemistry lab report. She has made four independent measurements of the specific gravity of a compound. The results are

$$3.82 \quad 3.93 \quad 3.67 \quad 3.78$$

The lab manual says that repeated measurements will vary according to a normal distribution with standard deviation $\sigma = 0.15$. (This standard deviation shows how precise the measurement process is.) The mean μ of the distribution of measurements is the true specific gravity.

(a) The lab manual asks for a 95% confidence interval for the true specific gravity. Your friend doesn't know how to do this. Do it for her.

(b) Now explain to your friend in simple language what "95% confidence" means.

(c) What critical value from the normal table would you use if you wanted 80% confidence instead of 95% confidence? Would the 80% confidence interval be wider or narrower than your 95% confidence from (a)? [Do *not* actually compute the 80% confidence interval.]

(d) The lab manual also asks whether the data show convincingly that the true specific gravity is less than 3.9. State the null and alternative hypotheses used to answer this question. Then calculate the test statistic, and find its P-value.

(e) Explain to your friend in one or two sentences what the specific P-value you found in (c) means.

4.4   Sample examination II solutions

1. (a) First (Rule of thumb 1), the population of California is much larger than 10 times the sample size $n = 1000$. So we can use the usual formula for the standard deviation of $\hat{p}$.

Second (Rule of thumb 2), for $n = 1000$ and $p = .22$ we have $np = 220$ and $n(1-p) = 780$. Both are much larger than 10, so the normal approximation will be quite accurate.

(b) The question concerns the *count* of 1000 California residents born outside the U.S. Translate into a question about the *proportion*,

$$\text{count} \geq 250 \quad \text{is} \quad \hat{p} = \frac{\text{count}}{1000} \geq \frac{250}{1000} = .25$$

The mean of $\hat{p}$ is $p = .22$. The standard deviation is

$$\sqrt{\frac{p(1-p)}{n}} = \sqrt{\frac{(.22)(.78)}{1000}} = .0131$$

Now use the normal approximation to find the probability:

$$\begin{aligned} P(\hat{p} \geq .25) &= P\left(\frac{\hat{p} - .22}{.0131} \geq \frac{.25 - .22}{.0131}\right) \\ &= P(Z \geq 2.29) \\ &= 1 - .9890 = .0110 \end{aligned}$$

2. (a) The weight x of a single child has the $N(7.5, 1.25)$ distribution. So

$$\begin{aligned} P(x < 5.5) &= P\left(\frac{x - 7.5}{1.25} < \frac{5.5 - 7.5}{1.25}\right) \\ &= P(Z < -1.60) = .0548 \end{aligned}$$

(b) The mean birth weight $\bar{x}$ of a sample of 3 children still has mean 7.5 pounds, but its standard deviation is

$$\frac{\sigma}{\sqrt{3}} = \frac{1.25}{\sqrt{3}} = .7217 \text{ pound}$$

The probability we want is therefore

$$\begin{aligned} P(\bar{x} < 5.5) &= P\left(\frac{\bar{x} - 7.5}{.7217} < \frac{5.5 - 7.5}{.7217}\right) \\ &= P(Z < -2.77) = .0028 \end{aligned}$$

3. (a) Use Table C to see that the upper 0.025 critical value is $z^* = 1.960$.

   (b) The key words are "is different." The alternative hypothesis is two-sided,

$$H_0 : \mu_G = \mu_B$$

$$H_a : \mu_G \neq \mu_B$$

Here μ_G and μ_B are the mean milk production for all cows of this breed with forage spread on the ground and in bunks, respectively. We do not yet know how to test these hypotheses.

(c) No. The margin of error in a confidence interval covers only the random sampling error due to chance variation in random sampling.

(d) The standard deviation goes down as the sample size n goes up, at the rate $\sqrt{n}$. (The standard deviation of a sample proportion $\hat{p}$ is $\sqrt{p(1-p)/n}$.) So a sample 9 times larger has a standard deviation one-third as large.

(e) A result significant at the 5% level is in the extreme 5% of the sampling distribution; so it may or may not also be in the extreme 1%.

4. (a) The sample mean is $\bar{x} = 3.80$. The 95% confidence interval is therefore

$$
\begin{aligned}
\bar{x} \pm z^{*} \frac{\sigma}{\sqrt{n}} &= 3.80 \pm 1.960 \frac{.15}{\sqrt{4}} \\
&= 3.80 \pm .147
\end{aligned}
$$

(b) "95% confidence" means that we got this interval by using a method that in 95% of all samples will produce an interval that covers the true specific gravity.

(c) $z^{*} = 1.282$, the upper 0.10 critical value. The 80% confidence interval is narrower than the 95% confidence interval because the critical value required is smaller.

(d) The hypotheses are

$$H_0 : \mu = 3.9$$
$$H_a : \mu < 3.9$$

The test statistic is

$$
\begin{aligned}
z = \frac{\bar{x} - \mu_0}{\sigma/\sqrt{n}} &= \frac{3.8 - 3.9}{.15/\sqrt{4}} \\
&= -1.33
\end{aligned}
$$

and its P-value (one-sided on the low side) is

$$P(Z \leq -1.33) = .0918$$

(e) There is probability 0.0918 that the mean of 4 readings would be as small as 3.8 if the true specific gravity were 3.9. That is, we observed a value in the smallest 9.2% of all results we could get if 3.9 were correct. This is only weak evidence that the specific gravity is less than 3.9, because a value this small would come up more than 9% of the time just by chance.

4.5 Sample final examination

(15) 1. An historian examining British colonial records for the Gold Coast in Africa suspects that the death rate was higher among African miners than among

European miners. In the year 1936, there were 223 deaths among 33,809 African miners and 7 deaths among 1541 European miners in the Gold Coast. (Data courtesy of Raymond Dumett, Department of History, Purdue University.)

Consider this year as a sample from the pre-war era in Africa. Is there good evidence that the proportion of African miners who died during a year was higher than the proportion of European miners who died? [State hypotheses, calculate a test statistic, give a P-value as exact as the tables in the text allow, and state your conclusion in words.]

(25) 2. An agricultural researcher reasons as follows: A heavy application of potassium fertilizer to grasslands in the spring seems to cause lush early growth but depletes the potassium before the growing season ends. So spreading the same amount of potassium over the growing season might increase yields. He therefore compares two treatments: 100 pounds per acre of potassium in the spring (Treatment 1) and 50, 25, and 25 pounds per acre applied in the spring, early summer, and late summer (Treatment 2). The experiment is continued over several years because growing conditions may vary from year to year.

The table below gives the yields, in pounds of dry matter per acre. It is known from long experience that yields vary roughly normally. (Data from R. R. Robinson, C. L. Rhykerd, and C. F. Gross, "Potassium uptake by orchardgrass as affected by time, frequency and rate of potassium fertilization," *Agronomy Journal*, 54(1962) 351–353.)

Treatment	Year 1	Year 2	Year 3	Year 4	Year 5
1	3902	4281	5135	5350	5746
2	3970	4271	5440	5490	6028

(a) Do the data give good evidence that Treatment 2 leads to higher average yields? [State hypotheses, carry out a test, give a P-value as exact as the tables in the text allow, and state your conclusions in words.]

(b) Give a 98% confidence interval for the mean increase in yield due to spreading potassium applications over the growing season.

(15) 3. Prior to an intensive TV advertising campaign, the producers of Nike athletic shoes find that 29 of a random sample of 200 upper-income adults are aware of their new leisure shoe line. A second random sample of 300 such adults is taken after the campaign. Now 96 of the persons sampled can identify the new line.

Give a 99% confidence interval for the increase in the proportion of upper income adults showing brand awareness.

(30) 4. Here are data on the years of schooling completed x, and annual income, y, (in thousands of dollars) for a sample of 18 40-year old men.

Years	10	16	12	6	12	12	16	16	18
Income	28	38	16	13	25	30	35	27	28

Years	12	10	12	16	14	11	12	19	16
Income	28	26	21	34	30	21	27	29	24

A scatterplot (don't do it) shows a generally linear relation, but with considerable scatter about the line of best fit. A computer least squares regression program gives the output below. (The "Coeff" column gives the slope a and intercept b; the "Std Err" column gives the standard errors of these statistics. The "Residual Standard Error" is the observed standard deviation s about the regression line.)

```
                Coef      Std Err     t Value
Intercept     10.84249   5.103363    2.124577
x2             1.186813  0.372311    3.187693

Residual Standard Error = 5.02275     R-Square = 0.3884116
N = 18          F Value = 10.16139 on 1, 16 df
```

(a) What percent of the observed variation in income is explained by the straight line relation between income and education?

(b) Is there strong evidence that there is a straight line relation between education and income? (State hypotheses, carry out a test, use a table to find values between which the P-value falls, and state your conclusion.)

(c) Consider 40-year old men who have 16 years of education. (These are men with four years of college but no further education.) Give a 95% interval for their average income.

(15) 5. Answer each of the following questions. (No explanation is needed – just a short answer.)

(a) You are reading an article in your field that reports several statistical analyses. The article says that the P-value for a significance test is 0.045. Is this result significant at the 5% significance level?

(b) Is the result with P-value 0.045 significant at the 1% significance level?

(c) For another significance test, the article says only that the result was significant at the 1% level. Are such results always, sometimes, or never significant at the 5% level?

(d) The reaction times of a subject to a stimulus are strongly skewed to the right because of a few slow reaction times. You wish to test

$$H_0 : \mu_1 = \mu_2$$

where μ_1 is the mean reaction time for Stimulus 1, and μ_2 for Stimulus 2. You have two independent samples, 8 observations for Stimulus 1 and 10 for Stimulus 2. Which, if any, of the tests you have studied can be used to test this?

(e) You read an article that contains a 95% confidence interval. Would the margin of error in a 99% confidence interval computed from the same data be less, the same, or greater?

4.6   Sample final examination solutions

1. This is a two-sample setting, with

> Population 1 = African miners
> Population 2 = European miners

We want to test

$$H_0 : p_1 = p_2$$
$$H_a : p_1 > p_2$$

The two sample proportions are

$$\hat{p}_1 = \frac{223}{33,809} = .006596 \quad \text{and} \quad \hat{p}_2 = \frac{7}{1541} = .004543$$

The pooled sample proportion is therefore

$$
\begin{aligned}
\hat{p} &= \frac{\text{count of deaths in both samples combined}}{\text{count of miners in both samples combined}} \\
&= \frac{223 + 7}{33,809 + 1541} \\
&= \frac{230}{35,350} = .006506
\end{aligned}
$$

and the z test statistic is

$$
\begin{aligned}
z &= \frac{\hat{p}_1 - \hat{p}_2}{\sqrt{\hat{p}(1 - \hat{p})\left(\frac{1}{n_1} + \frac{1}{n_2}\right)}} \\
&= \frac{.006596 - .004543}{\sqrt{(.006506)(.993494)\left(\frac{1}{33,809} + \frac{1}{1541}\right)}} \\
&= \frac{.002053}{.0020943} = .980
\end{aligned}
$$

Table A gives the P-value as $1 - .8365 = .1635$. There is, surprisingly, no significant evidence that the African death rate is higher.

2. This is a *matched pairs* setting because the observations are paired by years.

    (a) The hypotheses, expressed in terms of the mean differences, Treatment 2 − Treatment 1, are

$$H_0 : \mu = 0$$
$$H_a : \mu > 0$$

The differences are

$$68 \quad -10 \quad 305 \quad 140 \quad 282$$

with

$$\bar{x} = 157 \quad \text{and} \quad s = 135.672$$

Apply the one-sample t test to these differences. The test statistic is

$$t = \frac{\bar{x} - 0}{s/\sqrt{n}} = \frac{157}{135.672/\sqrt{5}} = 2.588$$

The P-value based on the t distribution with $n - 1 = 4$ degrees of freedom falls between 0.025 and 0.05 (using table C). This is moderately strong evidence that Treatment 2 produces a higher mean yield.

    (b) For 98% confidence and 4 degrees of freedom, use $t^* = 3.747$. The confidence interval is

$$\begin{aligned}
\bar{x} \pm t^* \frac{s}{\sqrt{n}} &= 157 \pm 3.747 \frac{135.672}{\sqrt{5}} \\
&= 157 \pm 227.3 \\
&= (-70.3, \ 384.3)
\end{aligned}$$

3. There are two independent samples. We want a confidence interval for a difference between two population proportions. The sample proportions are

$$\hat{p}_1 = \frac{29}{200} = .145 \quad \text{and} \quad \hat{p}_2 = \frac{96}{300} = .320$$

We can use procedures based on the normal approximation because the population is large and

$$n\hat{p}_1 = 29 \quad n(1 - \hat{p}_1) = 191 \quad n\hat{p}_2 = 96 \quad n(1 - \hat{p}_2) = 204$$

are all more than 5. The standard error for $\hat{p}_2 - \hat{p}_1$ is

$$\begin{aligned}
\text{SE} &= \sqrt{\frac{\hat{p}_1(1 - \hat{p}_1)}{n_1} + \frac{\hat{p}_2(1 - \hat{p}_2)}{n_2}} \\
&= \sqrt{\frac{(.145)(.855)}{200} + \frac{(.320)(.680)}{300}} \\
&= \sqrt{.0013452} = .03668
\end{aligned}$$

The 99% confidence interval for $p_2 - p_1$ is

$$
\begin{aligned}
(\hat{p}_2 - \hat{p}_1) \pm z^*\text{SE} &= (.320 - .145) \pm (2.576)(.03668) \\
&= .175 \pm .0945 \\
&= (.0805, .2695)
\end{aligned}
$$

We are 99% confident that between 8% and 27% of upper-income adults are aware of the new shoe line.

4. (a) The output says R-Square = 0.3884116, so the linear relationship explains 38.8% of the observed variation in income.

   (b) The null hypothesis of "no relation" says that the slope of the true regression line is 0. The hypotheses are:

$$H_0 : \beta = 0$$

$$H_a : \beta \neq 0$$

   The computer output shows that the t statistic for the test is $t = 3.187693$. The degrees of freedom are $n - 2 = 16$. From Table C we see that t falls between the 0.0025 and 0.005 upper critical values of $t(16)$. Doubling these values because H_a is two-sided, $0.005 < P < 0.01$. There is strong evidence that a linear relationship exists.

   (c) The predicted mean income for $x = 16$ is

$$\hat{y} = 10.84249 + (1.186813)(16) = 29.831$$

   or \$29,831. The rest of this is a bit tedious by hand, so consider your options. Here goes. Using a calculator for $\bar{x}$ and s_x gives that $\bar{x} = 13.33$ and

$$\sum (x - \bar{x})^2 = (n - 1)s_x^2 = (17)(3.27198)^2 = 182$$

   The proper standard error for estimating the mean income is

$$
\begin{aligned}
\text{SE}_{\hat{\mu}} &= s\sqrt{\frac{1}{n} + \frac{(x^* - \bar{x})^2}{\sum (x - \bar{x})^2}} \\
&= 5.02275\sqrt{\frac{1}{18} + \frac{(16 - 13.33)^2}{182}} \\
&= 5.02275\sqrt{.0946275} = 1.5451
\end{aligned}
$$

   The 90% confidence interval is therefore

$$
\begin{aligned}
\hat{y} \pm t^*\text{SE}_{\hat{\mu}} &= 29.831 \pm (2.120)(1.5451) \\
&= 29.831 \pm 3.276 \\
&= (26.555, 33.107)
\end{aligned}
$$

or \$26,555 to \$33,107.

5. (a) Yes. The P-value is less than 0.05.

    (b) No. The P-value is greater than 0.01.

    (c) A result significant at the 1% level lies in the extreme 1% of a sampling distribution. This is certainly in the extreme 5%, so the result is always significant at the 5% level.

    (d) The samples are small and the distributions are strongly skewed. It would be unwise to use a t test in this setting. None of the tests we have studied is appropriate.

    (e) The margin of error would be greater. Higher confidence is paid for with a greater margin of error.

SOLUTIONS TO EXERCISES

By Professor Darryl Nestor
Bluffton College
Bluffton, Ohio

CHAPTER 1 SOLUTIONS

Section 1.1

1.1 "Region" is categorical; all others are quantitative.

1.2 (a) categorical; **(b)** quantitative; **(c)** categorical; **(d)** categorical; **(e)** quantitative; **(f)** quantitative.

1.3 (a) Roughly symmetric, though it might be viewed as SLIGHTLY skewed to the right. **(b)** About 15%. (39% of the stocks had a total return less than 10%, while 60% had a return less than 20%. This places the center of the distribution somewhere between 10% and 20%.) **(c)** The smallest return was between −70% and −60%, while the largest was between 100% and 110%. **(d)** 23% (1+1+1+1+3+5+11).

1.4 (a) Skewed to the right; center at about 3 (31 less than 3, 11 equal to 3, 23 more than 3); spread: 0 to 10. No outliers. **(b)** about 23% (15 out of 65 years).

1.5 Lightning histogram: centered at noon (or more accurately, somewhere from 11:30 to 12:30). Spread is from 7 to 17 (or more accurately, 6:30am to 17:30, i.e., 5:30pm). Shakespeare histogram: centered at 4, spread from 1 to 12.

1.6 (a) Table at right. **(b)** Histogram below. Children (under 10) represent the single largest group in the population; about one out of five Americans was under 10 in 1950. There is a slight dip in the 10–19 bracket, then the percentages trail off gradually after that. **(c)** Histogram below. The projections show a much greater proportion in the higher age brackets—there is now a gradual rise in the proportion up to ages 40–49, followed by the expected decline in the proportion of "senior citizens."

Age Group	1950	2075
0–9	19.4%	11.2%
10–19	14.4	11.5
20–29	15.9	11.8
30–39	15.1	12.3
40–49	12.8	12.2
50–59	10.3	12.1
60–69	7.3	11.1
70–79	3.6	8.8
80–89	1.1	6.1
90–99	0.1	2.5
100–109	0.0	0.5

(b)   **(c)**

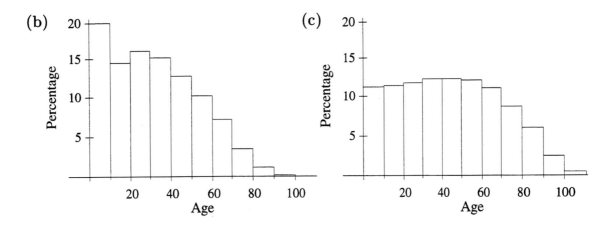

1.7 Outlier: 200. Center: 137 to 140 (there are nine observations less than or equal to 137, and nine greater than or equal to 140). Spread (ignoring the outlier): 101 to 178.

```
10 | 139
11 | 5
12 | 669
13 | 77
14 | 08
15 | 244
16 | 55
17 | 8
18 |
19 |
20 | 0
```

1.8 (a) Stemplot at right. **(b)** Distribution is skewed to the left, centered between 25 and 29 (the 10th and 11th scores). One might consider 10 to be an outlier, and possibly 15 as well. **(c)** 27 (or any score between 25 and 29).

```
1 | 0
1 | 5
2 | 0023344
2 | 59
3 | 000112223
```

1.9 (a) Below. **(b)** Prices increased throughout the entire period. **(c)** Prices rose most quickly from 1979–80 (the steepest part of the graph—or, from the table, an increase of 9.8 over one year). The slowest rise occurred from about 1970–73, with a similar rate of increase in 1985–86.

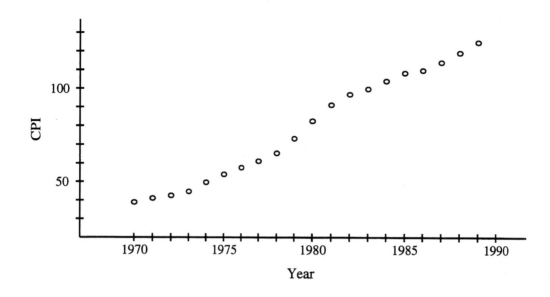

1.10 (a) The individuals are the company's employees. **(b)** The variables "Gender," "Race," and "Job Type" are categorical. **(c)** "Age" and "Salary" are quantitative, measured in years and dollars, respectively.

1.11 Slightly skewed to the right, centered at 4.

1.12 (a) Roughly symmetric (though with two apparent peaks). **(b)** The center (a "typical" batting average) is between .265 and .275, and the spread is from .185 to .355 (ignoring Brett).

1.13 See next problem for stemplot. The distribution is skewed to the left, and centered at 46. 60 is *not* an outlier.

1.14 Maris' 61 home-run year is an outlier—a clear departure from the other nine years. The bulk of Ruth's stemplot (on the right) lies (*physically*) below (and *numerically* above) Maris'. Ruth's four lowest years are at about the same level as Maris' best five years (excluding the outlier).

8	0	
43	1	
6	1	
3	2	2
86	2	5
3	3	4
9	3	5
	4	11
	4	66679
	5	44
	5	9
1	6	0

1.15 (a) After rounding all numbers to the first place after the decimal and splitting stems, the plot shows the distribution to be fairly symmetric, with a high outlier of 4.7 (4.69). **(b)** The time plot shows no clear trend.

1	12
1	556799
2	0111122233
2	568899
3	0113
3	5
4	
4	7

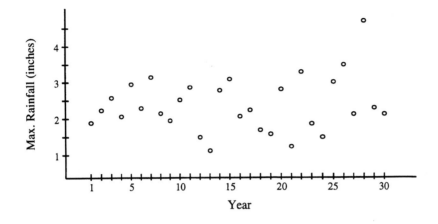

1.16 (a) Round to nearest integer before creating stemplot. **(b)** There is no particular observable shape (considering symmetry and skewness). **(c)** (Time plot below). **(d)** The time plot shows an increasing trend—adjustments should be made to counteract the rising tensions.

25	7
26	5
27	00
28	034
29	7
30	58
31	08
32	78
33	69
34	033
35	
36	
37	5

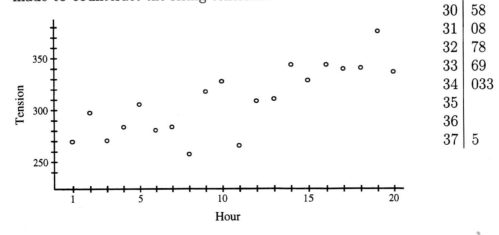

1.17 Shown on right is one possible plot with a compressed time axis, so that it appears to be steeper than Figure 1.7. Plots that appear in newspapers or magazines may be "distorted" (stretched or compressed) in this way so that they fit into a particular space.

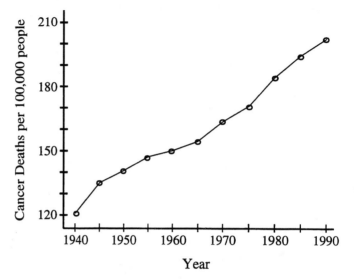

1.18 There are two distinct groups of states—"less than 30%" and "more than 40%." There are no particular outliers.

```
0 | 4
0 | 55566899
1 | 0001222234
1 | 567
2 | 024
2 | 58
3 |
3 |
4 | 2244
4 | 59
5 | 244
5 | 57889
6 | 0224
6 | 789
7 | 024
```

1.19 Stems are split *five* ways. The distribution seems to be skewed to the right.

```
2 | 23
2 | 4445
2 | 6666777
2 | 8888999999
3 | 01111
3 | 2222233333
3 | 55
3 | 66
3 | 8888
4 | 00
4 | 23
4 | 4
```

1.20 Skewed to the right. New Jersey (at \$9159 per student) might be considered outlier.

3	0233
3	67777
4	02234444
4	688889
5	0000112233334
5	56799
6	024
6	5
7	0
7	99
8	2
8	5
9	2

Section 1.2

1.21 (a) $\bar{x} = 2539 \div 18 = 141.058$. **(b)** After dropping the outlier, $\bar{x}^* = 2339 \div 17 = 137.588$. This is more in agreement with the "center" found in Ex. 1.7—the outlier makes the mean higher than it "should" be.

1.22 For Ruth: $M = 46$; for Maris: $M = 24.5$.

1.23 $M = 138.5$. The median is smaller than the mean (141.058), because with the outlier included, the distribution is skewed to the right.

1.24 $\bar{x} = \$480,000 \div 8 = \$60,000$. Seven of the eight employees (everyone but the owner) earned less than the mean. $M = \$22,000$.

1.25 The median is the smaller number (\$490,000)—the distribution is skewed to the right, which increases the mean but not the median.

1.26 (a) Ruth: 22 35 46 54 60. Maris: 8 14 24.5 33 61. **(b)** As in **1.14**, we see that for the most part, Ruth had better seasons than Maris.

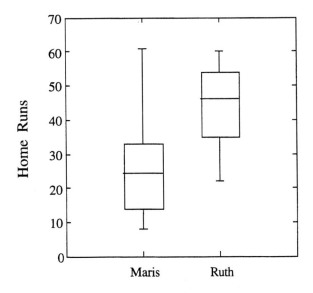

1.27 (a) It appears that M should be about the same as $\overline{x}$, as there is no particular skewness. **(b)** Five-number summary: 42 51 55 58 69. $\overline{x} = 54.833$, confirming our answer to (a). **(c)** Between Q_1 and Q_3: 51 to 58.

1.28 Both boxplots have approximately the same *shape*, although the math scores have a slightly greater range. There is a definite difference in *location*, however: math scores tend to be higher than verbal scores. The *lowest* math score is almost the same as the *median* verbal score.

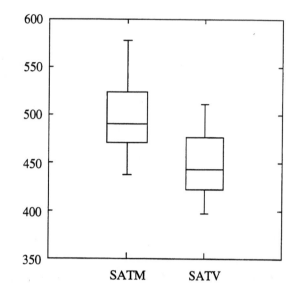

1.29 (a) $\overline{x} = 32.4 \div 6 = 5.4$. **(b)** $\sum(x_i - \overline{x})^2 = (0.2)^2 + (-0.2)^2 + (-0.8)^2 + (-0.5)^2 + (0.3)^2 + (1.0)^2 = 2.06$; $s^2 = 2.06 \div 5 = 0.412$; $s = \sqrt{0.412} = 0.6419$. **(c)** [Calculator work].

1.30 (a) Using $\overline{x} = 41.3$: $\sum(x_i - \overline{x})^2 = 5986.65$; $s^2 = 5986.65 \div 14 = 427.62$; $s = \sqrt{427.62} = 20.68$. **(b)** Calculators should give $s = 20.61$ (this results from using $\overline{x} = 41\frac{1}{3}$ instead of 41.3). Omitting the two outliers, we find $\overline{x} = 34.54$ and $s = 10.97$—both quantities had been increased by the skewness.

1.31 The stemplot reveals two peaks with a "valley" in between—one around 470 and one around 520. The mean and median fall between these two peaks.

43	7
44	013
45	9
46	113366
47	00013368
48	144667
49	079
50	2
51	1344799
52	1233578
53	9
54	2368
55	5
56	4
57	7

1.32 (a) Stemplot on right. **(b)** $M = 52.3$. **(c)** $Q_3 = 58.1$; there were landslides in 1964, 1972, and 1984.

```
4 | 33
4 |
5 | 00014
5 | 579
6 | 11
```

1.33 There seems to be little difference between beef and meat hot dogs, but poultry hot dogs are generally lower in calories than the other two. In particular, the median number of calories in a poultry hot dog is smaller than the lower quartiles of the other two, and the poultry lower quartile is less than the minimum calories for beef and meat.

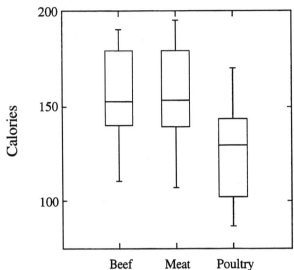

1.34 (a) From the MA and NE regions: 9159, 8500, 6534; and 7914, 5894, 6351, 5504, 6989, 5740. From ESC and SA (excluding DC): 3648, 4390, 3322, 3707; and 6016, 5154, 4860, 6184, 4802, 4327, 5360, 5046. **(b)** Northeastern: $\overline{x} = 6954$, $s = 1295$; five-number summary: 5504 5817 6534 8207 9159. Southern: $\overline{x} = 4735$, $s = 902$; five-number summary: 3322 3862 4831 5308 6184. (These are Minitab computations; computing by hand gives $Q_1 = 4017$ and $Q_3 = 5257$). The back-to-back stemplot (with southern states on the left and northeastern states on the right) shows the northeastern states are generally well ahead of the southern states in spending per pupil.

```
763 | 3 |
9843 | 4 | 579
 420 | 5 | 579
  20 | 6 | 45
     | 7 | 09
     | 8 | 5
     | 9 | 2
```

1.35 The stemplot shows the distribution to be fairly symmetrical, with a low outlier of 4.88—$\overline{x}$ and s should be reasonable in this setting. $\overline{x} = 5.4479$ and $s = 0.22095$; the mean $\overline{x}$ serves as our best estimate of the earth's density.

```
48 | 8
49 |
50 | 7
51 | 0
52 | 6799
53 | 04469
54 | 2467
55 | 03578
56 | 12358
57 | 59
58 | 5
```

1.36 Since there are two definite outliers (Alaska and Florida), the five-number summary is preferable; it is 4.2 11.35 12.65 13.7 18.3 (if computed by hand, $Q_1 = 11.4$). For reference, $\overline{x} = 12.544$ and $s = 2.121$.

1.37 The distribution is clearly skewed to the right, with at least the top two salaries (and arguably the top three) as outliers. The five-number summary is appropriate: 109 158 635 2300 6200.

```
0 | 111111111224
0 | 5668
1 | 0
1 | 5
2 | 1133
2 |
3 | 03
3 |
4 | 0
4 |
5 |
5 | 9
6 | 2
```

1.38 The difference in the mean and median indicates that the distribution of awards is skewed sharply to the right—i.e., there are some *very* large awards.

1.39 (a) Mean—although incomes are likely to be right-skewed, the city government wants to know about the total tax base. **(b)** Median—the sociologist is interested in a "typical" family, and wants to lessen the impact of the extremes.

1.40 The median—half are traveling faster than you, and half are traveling slower. (Actually, you have found *a* median—it could be that a whole range of speeds, say from 56 to 58 mph, might satisfy this condition.)

1.41 (a) 1, 1, 1, 1. **(b)** 0, 0, 10, 10. **(c)** For (a), any set of four identical numbers will have $s = 0$. For (b), the answer is unique; here is a rough description of why. We want to maximize the "spread-out"-ness of the numbers (that is what standard deviation measures), so 0 and 10 seem to be reasonable choices based on that idea. We also want to make each individual squared deviation—$(x_1 - \overline{x})^2$, $(x_2 - \overline{x})^2$, $(x_3 - \overline{x})^2$, and $(x_4 - \overline{x})^2$—as large as possible. If we choose 0, 10, 10, 10—or 10, 0, 0, 0—we make the first squared deviation (7.5^2), but the other three are only $(2.5)^2$. Our best choice is two at each extreme, which makes all four squared deviations equal to 5^2.

Section 1.3

1.42 There are many correct drawings. Here are two possibilities:

(a)   **(b)**

1.43 (a) 20% (the region is a rectangle with height 1 and base width 0.2; hence the area is 0.2). **(b)** 60%. **(c)** 50%.

1.44 (a) Mean is C, median is B. **(b)** Mean A, Median A; **(c)** Mean A, Median B.

1.45

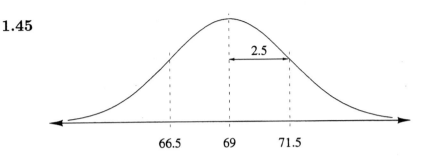

1.46 (a) 2.5% (this is 2 standard deviations above the mean). **(b)** 69 ± 5; that is, 64 to 74 inches. **(c)** 16%.

1.47 (a) 50%. **(b)** 2.5%. **(c)** 110 ± 50, or 60 to 160.

1.48 Eleanor's z-score is $(680 - 500)/100 = 1.8$; Gerald's is $(27 - 18)/6 = 1.5$. Eleanor's score is higher.

1.49 (a) 0.9978.   **(b)** $1 - 0.9978 = 0.0022$.

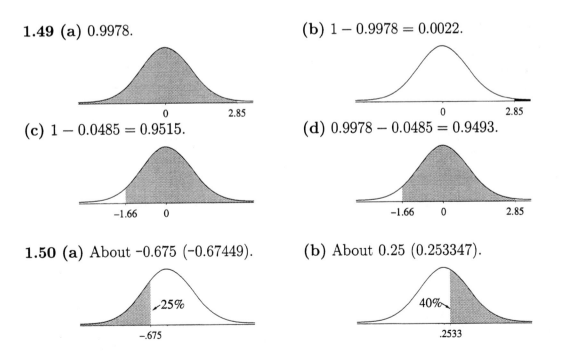

(c) $1 - 0.0485 = 0.9515$.   **(d)** $0.9978 - 0.0485 = 0.9493$.

1.50 (a) About −0.675 (−0.67449).   **(b)** About 0.25 (0.253347).

1.51 (a) $z = 3/2.5 = 1.20$; $1 - 0.8849 = 0.1151$; about 11.5%. **(b)** About 88.5%. **(c)** 72.2 inches.

1.52 (a) 65.5%. **(b)** 5.5%. **(c)** About 127 (or more).

1.53 $\bar{x} = 5.4479$ and $s = 0.22095$. About 75.8% (22 out of 29) lie within one standard deviation of $\bar{x}$, while 96.6% (28/29) lie within two standard deviations.

1 (3.4%)	2 (6.9%)	11 (37.9%)	11 (37.9%)	4 (13.8%)	0 (0%)

5.01 $\bar{x} - 2s$	5.23 $\bar{x} - s$	5.45 $\bar{x}$	5.67 $\bar{x} + s$	5.89 $\bar{x} + 2s$

1.54 Approximately 0.2 (for the tall one) and 0.5.

1.55 (a) About 16%. **(b)** About 68%.

1.56 (a) 266 ± 32, or 234 to 298 days. **(b)** Less than 234 days.

1.57 Cobb: $z = (.420 - .266) \div .0371 = 4.15$; Williams: $z = 4.26$; Brett: $z = 4.07$. Williams z-score is highest.

1.58 (a) 0.0122. **(b)** 0.9878. **(c)** 0.0384. **(d)** 0.9494.

1.59 (a) About 0.84. **(b)** About 0.385.

1.60 (a) -21.4% to 45%. **(b)** About 23.9% (23.89%). **(c)** About 21%.

1.61 (a) About 5.21%. **(b)** 44%. **(c)** 279 days or longer.

1.62 (a) At about ± 0.675. **(b)** For any normal distribution, the quartiles are ± 0.675 standard deviations from the mean; for human pregnancies, the quartiles are 266 ± 10.8, or 255.2 and 276.8.

1.63 (a) At about ± 1.28. **(b)** 64.5 ± 3.2, or 61.3 to 67.7.

Chapter Review

1.64 (a) Since a person cannot choose the day on which he or she has a heart attack, one would expect that all days are "equally likely"—no day is favored over any other. While there is *some* day-to-day variation, this does seem to be supported by the chart. **(b)** Monday through Thursday are fairly similar, but there is a pronounced peak on Friday, and lows on Saturday and Sunday. Patients do have some choice about when they leave the hospital, and many probably choose to leave on Friday, perhaps so that they can spend the weekend with the family. Additionally, many hospitals cut back on staffing over the weekend, and they may wish to discharge any patients who are ready to leave before then.

1.65 (a) Stemplot is symmetric with no *obvious* outliers (although 10.17 and 9.75 seem to be unusually high, and 6.75 is extraordinarily low). **(b)** Plot appears below; note outliers show up more clearly there. **(c)** $\overline{x} = 8.3628$ and $s = 0.4645$. **(d)** Between $\overline{x} - s$ and $\overline{x} + s$ (7.8983 to 8.8273): 25 (64.1%). Between $\overline{x} - 2s$ and $\overline{x} + 2s$ (7.4338 to 9.2918): 37 (94.9%). Between $\overline{x} - 3s$ and $\overline{x} + 3s$ (6.9693 to 9.7563): 39 (100%). These compare very nicely with the 68–95–99.7 rule.

```
 6 | 8
 7 | 44
 7 | 88888999
 8 | 01122333444
 8 | 555667777888
 9 | 000012
 9 | 8
10 | 2
```

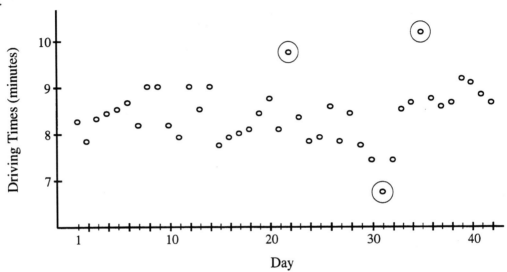

1.66 (a) Normal corn group: 272 333 (or 338) 358 401.2 (or 400.5) 462. New corn group: 318 379.25 (or 383.5) 406.5 429.25 (or 428.5) 477. The boxplot shows that the new corn seems to increase weight gain—in particular, the median weight gain for new-corn chicks was greater than Q_3 for those that ate normal corn. **(b)** Normal corn: $\overline{x} = 366.3$, $s = 50.8$; new corn: $\overline{x} = 402.95$, $s = 42.73$. On the average, the chicks which were fed the new corn gained 36.65 grams more mass (weight) than the other chicks.

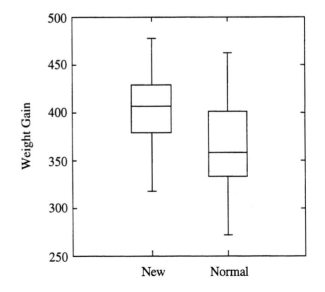

1.67 DiMaggio: 12 20.5 30 32 46. Mantle: 13 21 (or 20.5) 28.5 37 (or 37.75) 54. One might say that DiMaggio seems to have been more consistent—Mantle's plot is more spread out than DiMaggio's. The first three numbers in both summaries are similar, but Mantle's Q_3 and maximum are higher—he apparently had more impressive "big seasons" than DiMaggio.

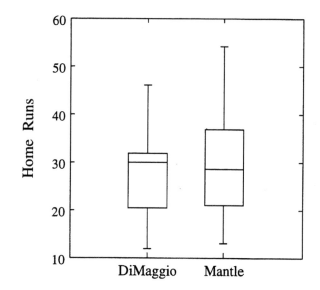

1.68 (a) Stems are split five ways. Distribution is fairly symmetric, except for the low outlier (15)—taking that into account, this might be considered left-skewed. **(b)** The Rolls Royce Silver Spur is certainly a gas-guzzler—and possibly the Mercedes S420. **(c)** 15 24.75 26 28 31 (if done by hand, $Q_1 = 25$). A typical car gets 26 MPG; the top quarter gets at least 28 MPG. **(d)** $\overline{x} = 25.8$—slightly lower than the median because of the low outlier (i.e., the left-skewedness).

```
1 | 5
1 |
1 |
2 | 0
2 | 233
2 | 445555
2 | 66666667777
2 | 888999
3 | 11
```

1.69 (a) Stems are the hundreds digit; leaves are tens digit. Clearly skewed to the right (as expected). Main peak occurs from 50 to 150—the guinea pigs which lived over 500 days are apparent outliers. **(b)** The mean is larger than the median because it is "drawn out" in the direction of the long tail—to the right. **(c)** 43 82.25 102.5 153.75 598. The difference between Q_3 and the maximum is relatively much larger than the other differences between successive numbers. This indicates a large "spread" among the high observations—that is, it shows that the data are skewed to the right.

```
0 | 4
0 | 55666677778888888888999999
1 | 000000000000011111222334444
1 | 556678889
2 | 0114
2 | 5
3 | 3
3 | 8
4 | 0
4 |
5 | 12
5 |
6 | 0
```

1.70 (a) -34.04 -2.95 3.47 8.45 58.68. **(b)** At end of best month: $1586.78; at end of worst month: $659.57. **(c)** Fairly symmetric—there are a several high and low outliers, but no particular skewness. (Note that the mean and median are quite similar.)

1.71 $IQR = 11.40$; subtracting $1.5 \times IQR$ from Q_1, and adding it to Q_3, gives the interval -20.05 to 25.55. Using this as a guide, the nine numbers listed as "Low" and "High" should be considered outliers.

1.72 (a) After the first two years, the median return is above zero all but once. However, there is no particular evidence of a trend. **(b)** The spread of the boxplots is considerably smaller in recent years (with the exception of 1987). **(c)** Four of the five high outliers are visible: 58.7 in 1973, 57.9 in 1975, 32 in 1979, and either 42 or 41.8 in 1974 (the fifth high outlier must have occured in one of the first three years, so that is it was overshadowed by a higher return). The lowest outlier appears in 1973, and the second lowest must have occurred there as well. In 1987, we see a return of either -26.6 or -27. These observations agree with the trend observed in (b)—most of the outliers occurred in the early years, and lately the variability has lessened considerably. The low in 1987 stands out as a "real" deviation from the pattern.

1.73 There is no apparent trend that would support either Julie's or John's position. However, one might observe a trend toward having a greater age spread in recent years (which, if true, is contrary to both positions—we do not simply have many older presidents, or many younger presidents, but rather we have a variety).

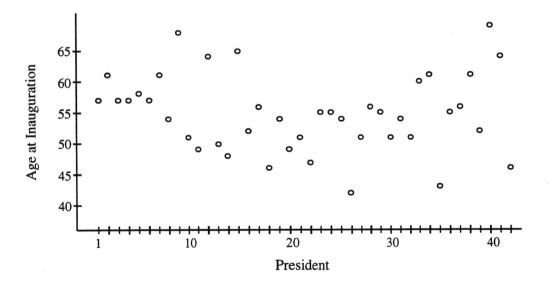

1.74 (a) Below. **(b)** The plot shows a decreasing trend—fewer disturbances overall in the later years—and more importantly, there is an apparent cyclic behavior. Looking at the table, the spring and summer months (April through September) generally have the most disturbances—probably for the simple reason that more people are outside during those periods.

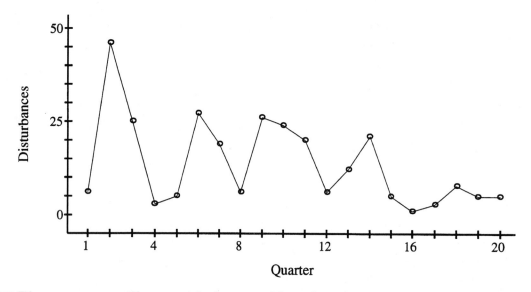

1.75 These answers will vary with the year. Note that the index of *Statistical Abstract* lists the *table* number, not the *page* number.

1.76 Total value of stock is likely to be skewed to the right—there are a (relatively) few companies with high market values which increase the mean, but not the median. (For example, Microsoft is listed on NASDAQ).

1.77 The proportion scoring below 1.7 is about 0.052; the proportion between 1.7 and 2.1 is about 0.078.

1.78 Soldiers whose head circumference is outside the range 22.8 ± 1.81—approximately, less than 21 in or greater than 24.6 in.

1.79 Those scoring at least 3.42 are in the "most Anglo/English" 30%; those scoring less than 2.58 make up the "most Mexican/Spanish" 30%.

CHAPTER 2 SOLUTIONS

2.1 Height at age six is explanatory, and height at age 16 is the response variable. Both are quantitative.

2.2 Sex is explanatory, and political preference in the last election is the response. Both are categorical.

2.3 "Treatment"—old or new—is the (categorical) explanatory variable. Survival time is the (quantitative) response variable.

Section 2.1

2.4 (a) Powerboat registrations is explanatory. **(b)** Plot shows that increased registrations seem to go along with more manatee deaths.

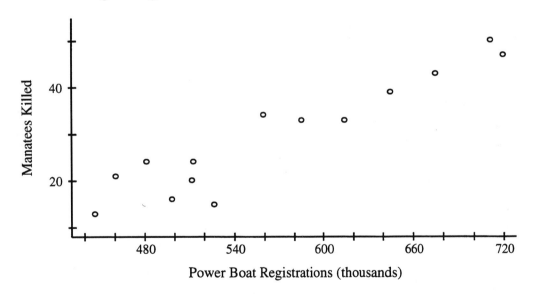

2.5 (a) Positive association. **(b)** Linear. **(c)** The relationship is fairly strong—it allows for reasonably accurate prediction. With 700,000 boat registrations, we would expect about 50 manatee deaths per year.

2.6 (a) Plot is below; speed is explanatory. **(b)** The relationship is curved—low in the middle, higher at the extremes. Since low "mileage" is actually *good* (it means that we use less fuel to travel 100 km), this makes sense: moderate speeds yield the best performance. Note that 60 km/hr is about 37 mph. **(c)** Above-average values of "mileage" are found with both low and high values of "speed." **(d)** The relationship is very strong—there is little scatter around the curve, and it is very useful for prediction.

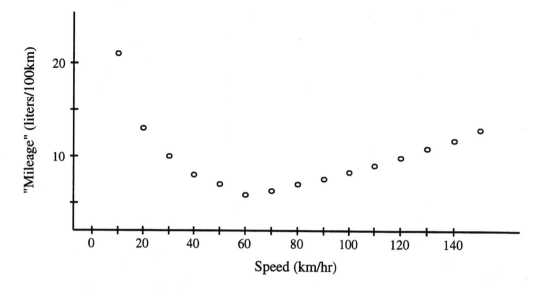

2.7 (a) See (c). Body mass is the explanatory variable. **(b)** Positive association, linear, moderately strong. **(c)** The male subjects' plot can be described in much the same way, though the scatter appears to be greater. The males typically have larger values for both variables.

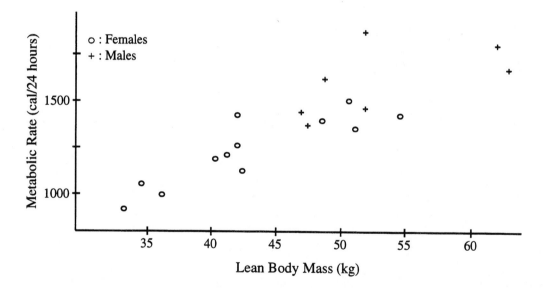

2.8 (a) Two mothers are 57 inches tall; their husbands are 66 and 67 inches tall. **(b)** The tallest fathers are 74 inches tall; there are three of them, and their wives are 62, 64 and 67 inches tall. **(c)** There is no clear explanatory variable; either could go on the horizontal axis. **(d)** The weak positive association indicates that people have *some* tendency to marry persons of a similar *relative* height—but it is not an overwhelming tendency. It is weak because there is a great deal of scatter.

2.9 (a) Lowest: about 107 calories (with about 145 mg of sodium); highest: about 195 calories, with about 510 mg of sodium. **(b)** There is a positive association; high-calorie hot dogs tend to be high in salt, and low-calorie hot dogs tend to have low sodium. **(c)** The lower left point is an outlier. Ignoring this point, the remaining points seem to fall roughly on a line. The relationship is moderately strong.

2.10 (a) At right. **(b)** Positive association; approximately linear save for two outliers (circled): spaghetti and the snack cake.

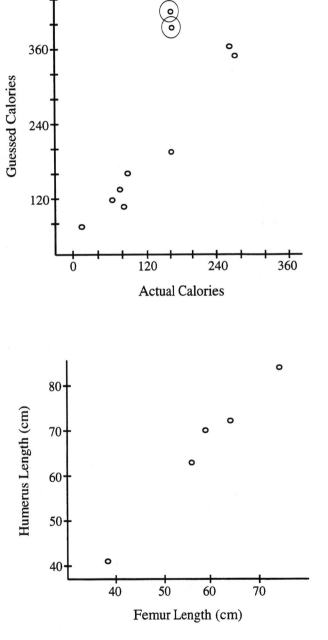

2.11 Since there is no obvious choice for response variable, either could go on the vertical axis. The plot shows a strong positive linear relationship, with no outliers. There appears to be only one species represented.

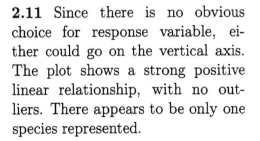

2.12 (a) Planting rate is explanatory. **(b)** See (d). **(c)** As we might expect from the discussion, the pattern is curved—high in the middle, and lower on the ends. Not linear, and there is neither positive nor negative association. **(d)** 20,000 plants per acre seems to give the highest average yield.

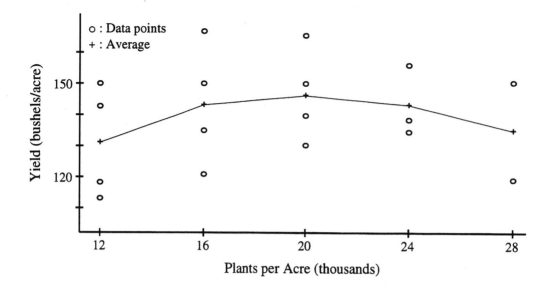

2.13 (a) There should be a positive association because money spent for teacher salaries is part of the education budget; more money spent per pupil would typically translate to more money spent overall. **(b)** See (e). **(c)** The plot shows a positive, approximately linear relationship. **(d)** California, spending $4826 per pupil, with median teacher salary $39,600. **(e)** The mountain states are clustered down in the lower left: they spend lower-than-average amounts per student, and have low median teacher salaries.

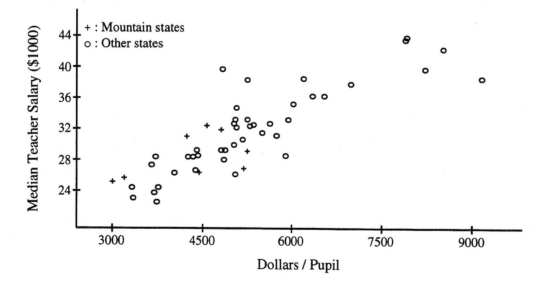

2.14 (a) There is a strong positive association—but it appears to be slightly curved rather than linear. **(b)** Now the strong positive association appears to be linear. (Note: the vertical axis scale may go from about 4.8 to 6.4, if the calculator used gives "natural" logarithms instead of "common (base 10)" logarithms.)

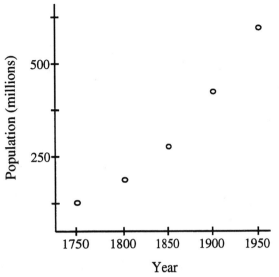

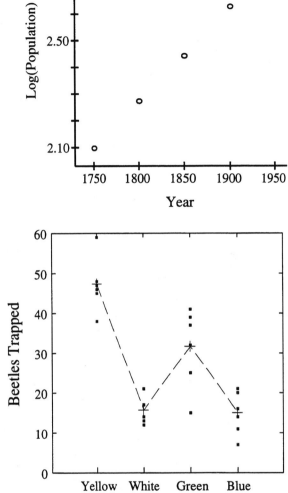

2.15 (a) Plot is on right. The means are (in the order given) 47.167, 15.667, 31.5, and 14.833. **(b)** Yellow seems to be the most attractive, and green is second. White and blue are poor attractors. **(c)** Positive or negative association make no sense here because color is a categorical variable (what is an "above-average" color?).

Section 2.2

2.16 In all cases, $\overline{x} = \overline{y} = 0$. **(a)** No association. $s_x = s_y = 4.6188$; $r = 0$. **(b)** Negative association. $s_x = 3.5355$, $s_y = 3.1623$; $r = -0.98387$. **(c)** Positive association. $s_x = s_y = 3.6515$; $r = 0.6$.

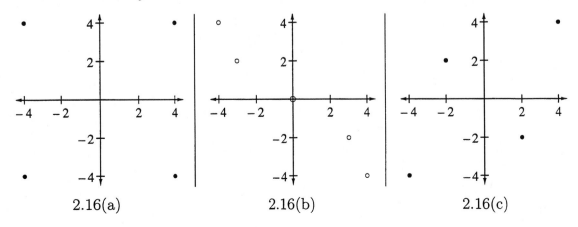

2.16(a)   2.16(b)   2.16(c)

2.17 **(a)** With x as femur length and y as humerus length: $\bar{x} = 58.2$, $s_x = 13.20$; $\bar{y} = 66.0$, $s_y = 15.89$; $r = 0.994$.

2.18 Because there is no obvious linear relationship, one expects the correlation to be near zero.

2.19 Clearly positive (there is positive association), but not near 1 (there is a fair amount of scatter).

2.20 $r = 1$.

2.21 **(a)** See problem 2.11 for plot. The plot shows a strong positive linear relationship, with little scatter, so we expect that r is close to 1. **(b)** r would not change—it is computed from standardized values, which have no units.

2.22 With x for speed and y for mileage: $\bar{x} = 40$, $s_x = 15.8$; $\bar{y} = 26.8$, $s_y = 2.68$; $r = 0$. Correlation only measures *linear* relationships; this plot shows a strong *non-linear* relationship.

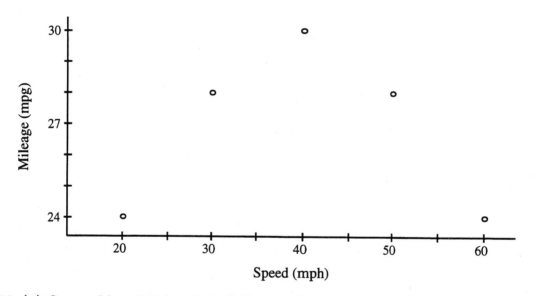

2.23 **(a)** See problem 2.7 for plot. Both correlations should be positive, but since the men's data seem to be more spread out, it may be slightly smaller. **(b)** Women: $r_w = 0.87645$; Men: $r_m = 0.59207$. **(c)** Women: $\bar{x}_w = 43.03$; Men: $\bar{x}_m = 53.10$. The difference in means has no effect on the correlation. **(d)** There would be no change, since standardized measurements are dimensionless.

2.24 **(a)** See problem 2.10 for plot. $r = 0.82450$. This agrees with the positive association observed in the plot; it is not too close to 1 because of the outliers. **(b)** It has no effect on the correlation. If every guess had been 100 calories higher— or 1000, or 1 million—the correlation would have been exactly the same, since the standardized values would be unchanged. **(c)** The revised correlation is $r = 0.98374$. The correlation got closer to 1 because without the outliers, the relationship is much stronger.

2.25 (a) The solid circles in the plot. **(b)** The open circles. **(c)** $r = 0.25310$ for both sets of data. For the first set, the x values range from -4 to 4, a spread of 8 units, while the spread in the y direction is only 1.1 units. For the second data set, the horizontal spread is small (only 0.8 units) compared to the vertical spread of 11 units. What matters in computing the correlation is not the "actual" sizes of the spreads in each direction (which is what we perceive in this plot), but rather the *relative* sizes of these spreads, which is more difficult to see unless we make two separate plots, each with appropriate x and y scales.

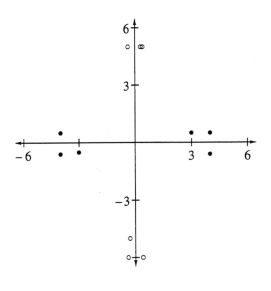

2.26 The person who wrote the article interpreted a correlation close to 0 as if it were a correlation close to -1. Prof. McDaniel's findings mean there is little linear association between research and teaching—for example, knowing a professor is a good researcher gives little information about whether she is a good or bad teacher.

2.27 The plot is given in problem 2.6. $r = -0.17162$—it is close to zero because the relationship is a curve rather than a line.

2.28 (a) Sex is a categorical variable. **(b)** r must be between -1 and 1. **(c)** r should have no units (i.e., it can't be 0.23 *bushel*).

Section 2.3

2.29 (a) $a = 1.0892$ and $b = 0.1890$, as given. **(b)** $\bar{x} = 22.31$, $s_x = 17.74$; $\bar{y} = 5.306$, $s_y = 3.368$; $r = 0.99526$. Except for roundoff error, we again find $b = 0.1890$ and $a = 1.0892$.

2.30 (a) A negative association—the pH decreased (i.e., the acidity increased) over the 150 weeks. **(b)** The initial pH was was 5.4247; the final pH was 4.6350. **(c)** The slope is -0.0053; the pH decreased by 0.0053 units per week (on the average).

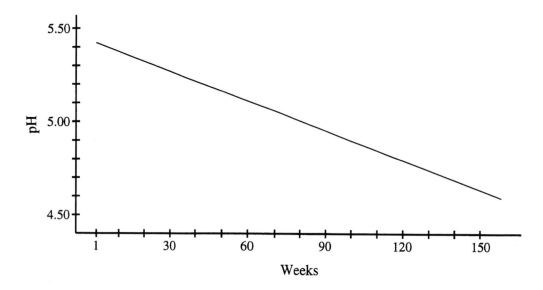

2.31 (a) Below. **(b)** The slope is close to 1—meaning that the strength after 28 days is *approximately* (strength after one week) plus 1389 psi. In other words, we expect the extra three weeks to add about 1400 psi of strength to the concrete. **(c)** 4557 psi.

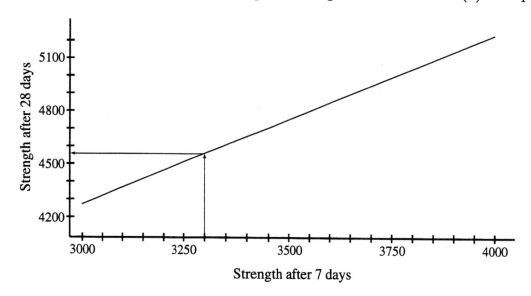

2.32 (a) $b = r \cdot s_y \div s_x = 0.16$; $a = \overline{y} - b\overline{x} = 30.2$. **(b)** $\hat{y} = 78.2$ **(c)** $r^2 = 0.36$; only 36% of the variability in y is accounted for by the regression, so the estimate $\hat{y} = 78.2$ could be quite different from the real score.

2.33 (a) Below. **(b)** There is a very strong positive linear relationship; $r = 0.9990$. **(c)** Regression line: $\hat{y} = 1.76608 + 0.080284x$ (y is steps/second, x is speed). **(d)** $r^2 = 0.998$, so nearly all the variation (99.8% of it) in steps taken per second is explained by the linear relationship. **(e)** The regression line would be different (as in Example 2.11), because the line in (c) is based on minimizing the sum of the squared *vertical* distances on the graph. This new regression would minimize the squared *horizontal* distances (for the graph shown). r^2 would remain the same, however.

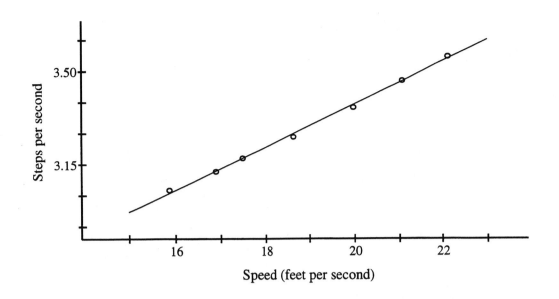

2.34 $r = \sqrt{0.16} = 0.40$ (high attendance goes with high grades, so the correlation must be positive).

2.35 **(a)** Below. **(b)** The line is clearly *not* a good predictor of the actual data—it is too high in the middle and too low on each end. **(c)** The sum is -0.01—a reasonable discrepancy allowing for round-off error. **(d)** Below.

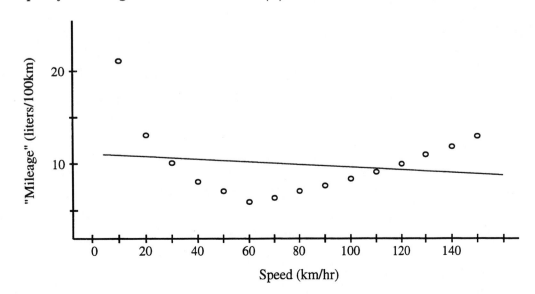

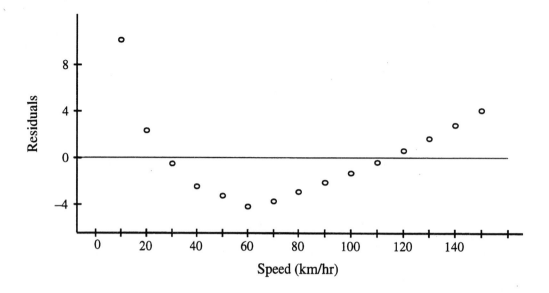

2.36 (a) At right. **(b)** Let y be "guessed calories" and x be actual calories. Using all points: $\hat{y} = 58.59 + 1.3036x$ (and $r^2 = 0.68$)—the dashed line. Excluding spaghetti and snack cake: $\hat{y}^* = 43.88 + 1.14721x$ (and $r^2 = 0.968$). **(c)** The two removed points could be called influential, in that when they are included, the regression line passes above every *other* point; after removing them, the new regression line passes throught the "middle" of the remaining points.

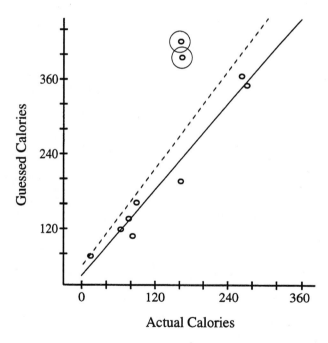

2.37 (a) Without Child 19, $\hat{y}^* = 109.305 - 1.1933x$. Child 19 might be considered *somewhat* influential, but removing this data point does not change the line substantially. **(b)** With all children, $r^2 = 0.410$; without Child 19, $r^2 = 0.572$. With Child 19's high Gesell score removed, there is less scatter around the regression line—more of the variation is explained by the regression.

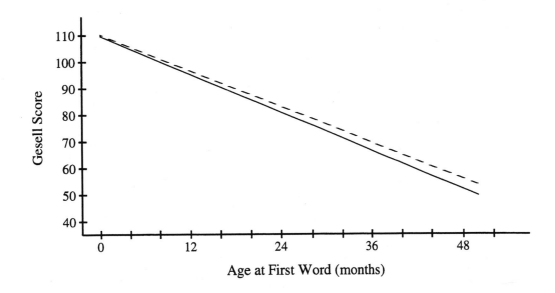

2.38 (a) Graph not shown. **(b)** $2500. **(c)** $y = 500 + 200x$.

2.39 (a) y (weight) $= 100 + 40x$ grams. **(b)** Graph not shown. **(c)** When $x = 104$, $y = 4260$ grams, or about 9.4 pounds—a rather frightening prospect. The regression line is only reliable for "young" rats; like humans, rats do not grow at a constant rate throughout their entire life.

2.40 Approximately 650 km/sec. Since there is quite a bit of variation around the line, and $r^2 = 0.615$, we should not have *too* much faith in the accuracy of this estimate.

2.41 (a) Below. **(b)** $\hat{y} = 71.950 + 0.38333x$. **(c)** When $x = 40$, $\hat{y} = 87.2832$; when $x = 60$, $\hat{y} = 94.9498$. **(d)** Sarah is growing at about 0.38 cm/month; she should be growing about 0.5 cm each month $(0.5 = \frac{6}{60-48})$.

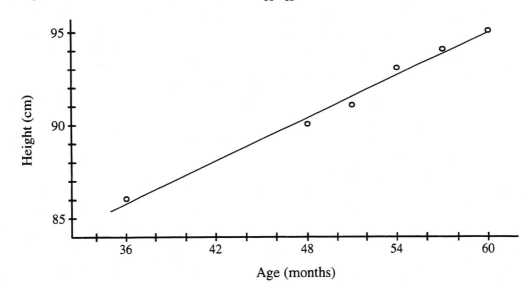

2.42 **(a)** Below. **(b)** $r = 0.56896$, and $r^2 = 0.32372$. There is a moderate positive association between US and overseas stock returns; this relationship allows us to explain about 32% of the variation in one quantity with the other. **(c)** The regression line is $\hat{y} = 4.777 + 0.8130x$. **(d)** $\hat{y} = 12.0\%$. We can only explain about one-third of the variation in overseas returns with the US return information, as evidenced by the wide scatter around the line, so we should not expect too much accuracy in our predictions. **(e)** The outlier point occurred in 1986. The two points on the left end of the graph—from 1973 and 1974—are potentially influential, especially the far left point.

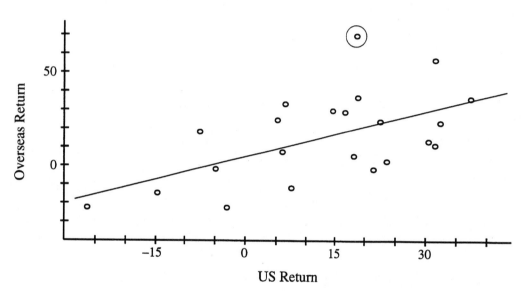

2.43 When $x = 480$, $\hat{y} = 255.95$ cm, or 100.77 in, or about 8.4 feet!

2.44 **(a)** US stocks: −26.4 3.05 17.5 25.3 37.2. Overseas stocks: −23.2 −2.0 15.45 30.35 69.40. **(b)** Overseas stocks generally had higher returns—as the five-number summaries and the boxplots show, a quarter of the time they did better than 30%. **(c)** The overseas stocks also fluctuated much more wildly—$(Q_3 - Q_1)$ is about 50% larger for overseas stocks, and the boxplot shows a lot more spread. Also, the low US return (−26.4%) is an outlier, while the overseas stocks have several negative values.

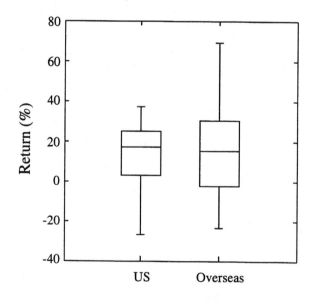

2.45 **(a)** About 69.4% of the variation is explained ($r^2 = 0.694$). **(b)** The sum is zero. **(c)** The residuals change from negative to positive in 1991—the year the change was made. In that year, the regression line changes from overestimating to underestimating.

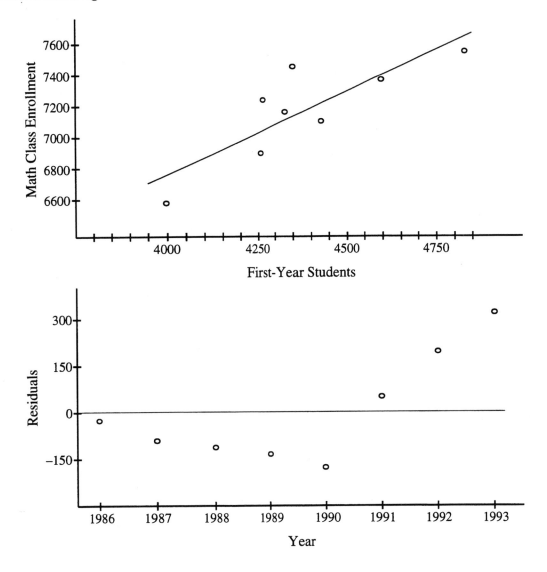

2.46 (a) The three correlations are $r_1 = 0.81642$, $r_2 = 0.81624$, and $r_3 = 0.81652$. The regressions yield: $\hat{y}_1 = 3.000 + 0.5001x$, $\hat{y}_2 = 3.001 + 0.5000x$, and $\hat{y}_3 = 3.002 + 0.4999x$. **(b)** Below. **(c)** For the first data set, the use of the regression line seems to be reasonable—the data do seem to have a moderate linear association (albeit with a fair amount of scatter). For the second, there is an obvious *non*-linear relationship— the regression line gives a poor estimate at $x = 14$. For the third, the data point with $x = 19$ is a very influential point—the other points alone give no indication of slope for the line. Seeing how widely scattered the y-coordinates of the other points are, we cannot place too much faith in the y-coordinate of the influential point; thus we cannot depend on the slope of the line, and so we cannot depend on the estimate when $x = 14$.

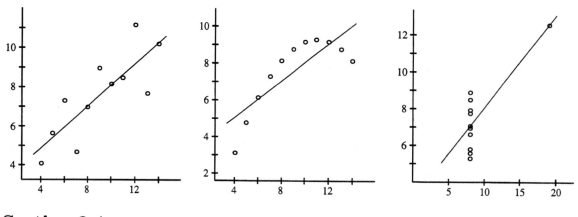

Section 2.4

2.47 **(a)** Regression line: $\hat{y} = 1166.93 - 0.58679x$. **(b)** Based on the slope, the farm population decreased about 590 thousand (0.59 million) people per year. The regression line explains 97.7% of the variation. **(c)** -782,100—clearly a ridiculous answer, since a population must be greater than or equal to 0.

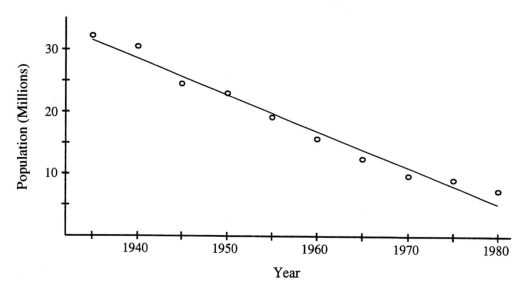

2.48 The explanatory and response variables were "consumption of herbal tea" and "cheerfulness." The most important lurking variable is social interaction—many of the nursing home residents may have been lonely before the students started visiting.

2.49 The correlation would be smaller because there is much more variation among the individual data points. This variation could not be as fully explained by the linear relationship between speed and step rate.

2.50 Seriousness of the fire is a lurking variable: more serious fires require more attention. It would be more accurate to say that a large fire "causes" more firefighters to be sent, rather than vice versa.

2.51 Age is the lurking variable here: we would expect both quantities—shoe size and reading level—to increase as a child ages.

2.52 No; more likely it means that patients with more serious conditions (which require longer stays) tend to go to larger hospitals, which are more likely to have the facilities to treat those problems.

2.53 (a) $r^2 = 0.925$—more than 90% of the variation in one SAT score can be explained through a linear relationship with the other score. (b) The correlation would be much smaller, since individual students have much more variation between their scores. Some may have greater verbal skills and low scores in math (or vice versa); some will be strong in both areas, and some will be weak in both areas. By averaging—or, as in this case, taking the median of—the scores of large groups of students, we muffle the effects of these individual variations.

2.54 The plot on the right is a very simplified (and not very realistic) example—open circles are economists in business; filled circles are teaching economists. The plot should show positive association when either set of circles is viewed separately, and should show a large number of bachelor's degree economists in business, and graduate degree economists in academia.

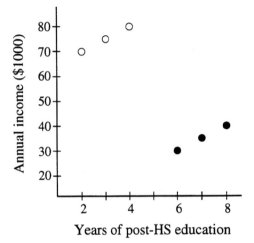

2.55 The explanatory variable is whether or not a student has taken at least two years of foreign language, and the score on the test is the response. The lurking variable is the students' English skills *before* taking (or not taking) the foreign language: students who have a good command of English early in their high school career are more likely to choose (or be advised to choose) to take a foreign language.

2.56 Social status is a possible lurking variable: children from upper-class families can more easily afford higher education, and they would typically have had better preparation for college as well. They may also have some advantages when seeking employment, and have more initial money should they want to start their own business.

    This could be compounded by racial distinctions: prejudicial hiring practices may keep minorities out of higher-paying positions.

    It also could be that some causation goes the other way: a man who is doing well in his job might be encouraged to pursue further education.

2.57 Apparently drivers are typically larger and heavier men than conductors—and are therefore more predisposed to health problems such as heart disease.

Section 2.5

2.58 Column sums to 38,663, which differs by 2 (thousand) from the total given. Roundoff error accounts for the difference.

2.59 27.0%, 24.4%, 16.2%, 13.5%, and 19.0% (total is 100.1% due to rounding).

2.60 (a) 5375. **(b)** $1004 \div 5375 = 18.7\%$. **(c)** Both parents smoke: 1780 (33.1%); One parent smokes: 2239 (41.7%); Neither parent smokes: 1356 (25.2%).

2.61 13.9%, 12.3%, 18.8%, 28.1%, and 42.2%. The percentage of people who did not finish high school increases fairly steadily with age.

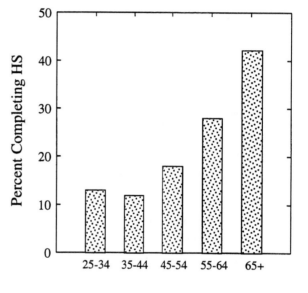

2.62 Compute $\dfrac{12702}{30092} = 42.2\%$, $\dfrac{10310}{30092} = 34.3\%$, etc.

2.63 Among 35 to 44 year-olds: 12.3% never finished high school, 37.5% finished high school, 22.7% had some college, and 27.5% completed college. This is more like the 25–34 age group than the 65 and over group.

2.64 Among those with 4 or more years of college: 29.9% are 25–34, 31.3% are 35–44, 17.5% are 45–54, 10.6% are 55–64, and 10.7% are 65 or older.

2.65 Two possible answers: Row 1–30, 20; Row 2–30, 20; and Row 1–10, 40; Row 2–50, 0.

2.66 (a) 6014; 1.26%. **(b)** Blood pressure is explanatory. **(c)** Yes: among those with low blood pressure, 0.785% died; the death rate in the high blood pressure group was 1.65%—about twice as high as the other group.

2.67 **(a)** Of students with two smoking parents, 22.5% smoke; with one smoking parent, the percentage drops to 18.6%; with no smoking parents, only 13.9% of the students smoke. **(b)** On right. **(c)** It appears that children of smokers are more likely to smoke— even more so when both parents are smokers.

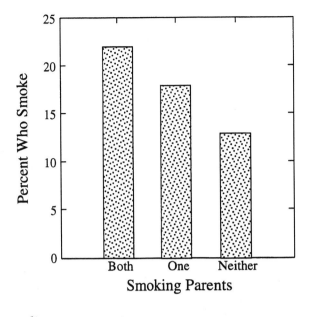

2.68 **(a)** On right. **(b)** 70% of male applicants are admitted, while only 56% of females are admitted. **(c)** 80% of male business school applicants are admitted, compared with 90% of females; in the law school, 10% of males are admitted, compared with 33.3% of females.

	Admit	Deny
Male	490	210
Female	280	220

(d) Six out of 7 men apply to the business school, which admits 83% of all applicants, while 3 of 5 women apply to the law school, which only admits 27.5% of its applicants.

2.69 **(a)** On right. **(b)** Overall, 11.9% of white defendants and 10.2% of black defendants get the death penalty. However, for white victims, the percentages are 12.6% and 17.5% (respectively); when the victim is black, they are 0% and 5.8%.

	Yes	No
White Defendant	19	141
Black Defendant	17	149

(c) In cases involving white victims, 14% of defendants got the death penalty; when the victim was black, only 5.4% of defendants were sentenced to death. White defendants killed whites 94.3% of the time—but are less likely to get the death penalty than blacks who killed whites.

2.70 **(a)** On right. **(b)** Joe: .240, Moe: .260. Moe has the best overall batting average. **(c)** Against right-handed pitchers: Joe: .400, Moe: .300. Against left-handed pitchers: Joe: .200, Moe: .100. Joe is better against both kinds of pitchers. **(d)** Both players do better against right-handed pitchers than against left-handed pitchers. Joe spent 80% of his at-bats facing lefties, while Moe only faced left-handers 20% of the time.

	Hit	No Hit
All pitchers		
Joe	120	380
Moe	130	370
Right-handed		
Joe	40	60
Moe	120	280
Left-Handed		
Joe	80	320
Moe	10	90

2.71 Apparently women are more likely to be in fields which pay less overall (to both men and women). For example, if many women, and few men, have Job A, where they earn $40,000 per year, and meanwhile few women and many men have Job B earning $50,000 per year, then lumping all women and men together leads to an incorrect perception of unfairness.

2.72 One very simple possibility is shown at right, using 10 smokers and 10 non-smokers. Lumped together, we find that there are 5 people in each classification (overweight/died early, etc.). There are, of course, infinitely many other examples.

Early Death?	Smoker Overweight? Yes	No	Non-Smoker Overweight? Yes	No
Yes	2	4	3	1
No	1	3	4	2

2.73 (a) 704,000. **(b)** 2,065,000. **(c)** Roundoff error.

2.74 (a) 11,374,000. **(b)** 51.2%. **(c)** 60.8%, 22.1%, 68.4%, and 11.0%. Bar chart at right. **(d)** The 18–21 age group makes up more than 60% of full-time students, but comprises less than 20% of part-time students.

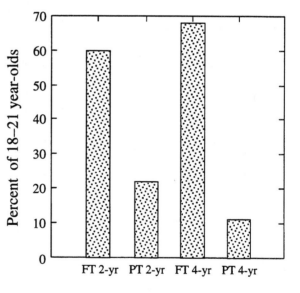

2.75 (a) 34.1%. **(b)** 36.9%.

2.76 (a) Counts: 127, 5829, 3031, 1907, 480. Percents: 1.1%, 51.2%, 26.6%, 16.8%, 4.2%. **(b)** 0.2%, 22.1%, 33.4%, 34.1%, and 10.1%. **(c)** The biggest difference between the distributions in (a) and (b) is that among part-time students at 2-year colleges, there is a markedly lower percentage of 18–21 year-olds, and considerable increases in the higher age brackets—the last two age categories are more than twice as large in (b) as they were in (a).

2.77 (a) 7.2%. **(b)** 10.1% of the restrained children were injured, compared to 15.4% of unrestrained children.

2.78 (a) 59.0%. **(b)** Larger businesses were less likely to respond: only 37.5% of the small businesses did not respond, compared to 59.5% of medium-sized businesses and 80% of large businesses. **(c)** Bar chart at right.

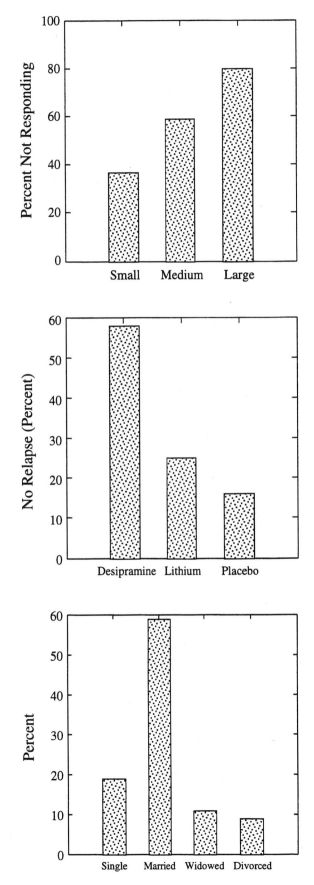

2.79 (a) 58.3% of desipramine users did not have a relapse, while 25.0% of lithium users and 16.7% of those who received placebos succeeded in breaking their addictions. **(b)** Because the addicts were assigned randomly to their treatments, we can *tentatively* assume causation (though there are other questions we need to consider before we can reach that conclusion).

2.80 (a) 12,625; roundoff error.
(b) 19.3%, 59.3%, 11.8%, and 9.6%.
(c) 18–24: 71.3%, 26.5%, 0.06%, 2.0%. 40–64: 5.8%, 72.5%, 7.6%, 14.1%. Among the younger women, almost three-fourths have not yet married, and those that are married have had little time to become widowed or divorced. Most of the older group is or has been married—only about 6% are still single. **(d)** 48.6% of single women are 18–24, 35.9% are 25–39, 10.7% are 40–64, and 4.9% are 65 or older.

Chapter Review

2.81 (a) Below. **(b)** There is no clear relationship. **(c)** Generally, those with short incubation periods are more likely to die. **(d)** Person 6—the 17-year-old with a short incubation (20 hours) who survived—merits extra attention. He or she is also the youngest in the group by far. Among the other survivors, one (person 17) had an incubation period of 28 hours, and the rest had incubation periods of 43 hours or more.

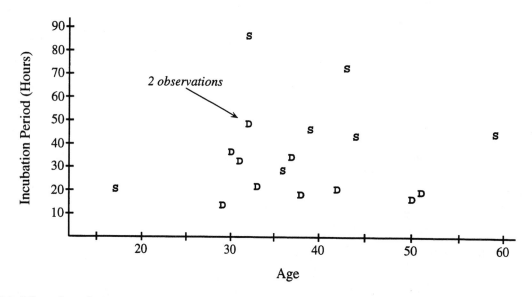

2.82 The plot shows an apparent negative association between nematode count and seedling growth. The correlation supports this: $r = -0.78067$. This also indicates that about 61% of the variation in growth can be accounted for by a linear relationship with nematode count.

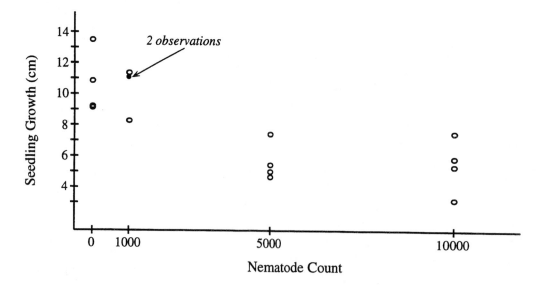

2.83 Both plots show a strong negative association—the logarithmic transformation is perhaps *slightly* more linear. A comparison of the correlation before and after the logarithm shows a slight increase (in absolute value): $r = -0.99178$ and $r_{\log} = -0.99684$.

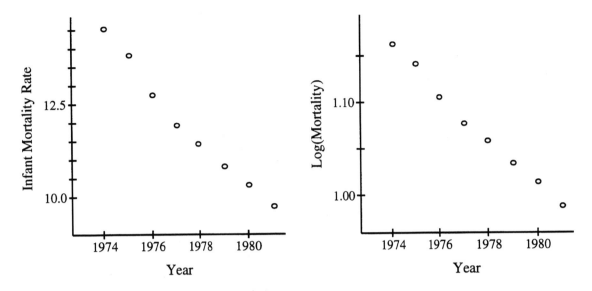

2.84 $b = 0.54$ (and $a = 33.67$). For $x = 67$ inches, we estimate $\hat{y} = 69.85$ inches.

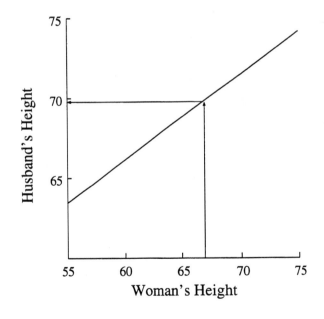

2.85 (a) The straight-line relationship is moderately strong; $r^2 = 88.6\%$. **(b)** When $x = 700$ (thousand), $\hat{y} = 46$. **(c)** Below. **(d)** -2.26 lies more than 2, but less than 3, standard deviations from the mean (0). This means that it is among the 5% most extreme values.

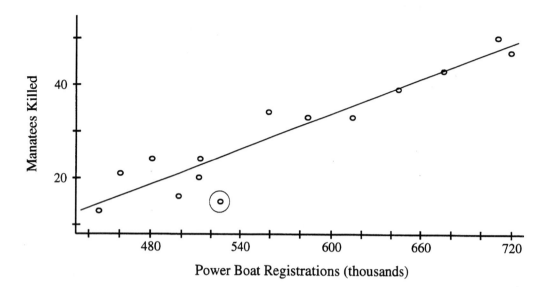

2.86 (a) Franklin is marked with a + (in the lower left corner). **(b)** There is a moderately strong positive linear association. (It turns out that $r^2 = 87.0\%$.) There are no really extreme observations, though Bank 9 did rather well. Franklin does not look out of place. **(c)** $\hat{y} = 7.573 + 4.9872x$. **(d)** Franklin's predicted income was $\hat{y} = 26.5$ million dollars—almost twice the actual income. The residual is –12.7.

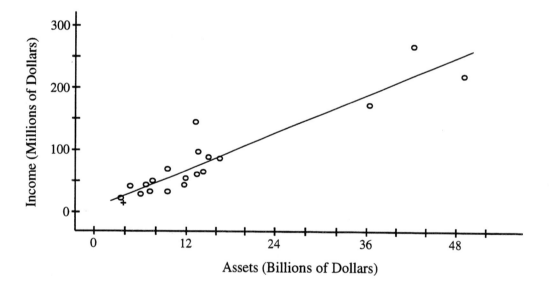

2.87 In the aspirin group, 1.26% had heart attacks (0.09% were fatal, 1.17% were not), and 1.08% had strokes. Among those who took placebos, 2.17% had heart attacks (0.24% fatal, 1.93% non-fatal), and 0.89% suffered from strokes. Based on these numbers, it appears that the aspirin group had a slight advantage in heart attacks, but was possibly worse in incidence of stroke. In spite of the word "study" in the name, this was an experiment, where the doctors could not choose for themselves which pill they took. Thus, a cause-and-effect relationship seems to be indicated—*but* we must be careful not to apply the results too broadly. The study involved *healthy,*

male doctors over 40; the same outcomes might not be observed for (e.g.) a person who already has heart problems, or a woman, or a patient under age 40.

2.88 80% of all suicide victims were men—that in itself is a major difference. Firearms were most the common method for both sexes: 65.9% of male suicides used firearms, as did 42.1% of females. Poison was a close second for women at 35.6%, compared with only 13.0% for men. In the last two categories, the percentages were pretty close: Men chose hanging 14.9% of the time, and women chose it 12.2% of the time; 6.2% of men and 10.1% of women fell into the "other" group.

2.89 (a) Filled circles are midsize cars; open circles are large. The squares represent 2 or more cars with the same city/highway mileage; the '+' at (18,25) represents one of each size (specifically, the Ford Crown Victoria and Lincoln Mark VIII).

The plot shows a positive linear association between the two variables—meaning, for example, that cars with above-average city mileage also do well on the highway. There is some evidence that the points for large cars appear higher on the graph, indicating that for a given city mileage, large cars tend to have a slightly higher highway

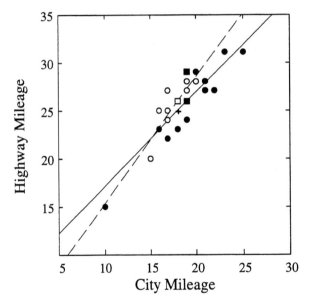

mileage. **(b)** The Rolls Royce Silver Spur has unusually low mileage (city and highway). **(c)** The relationship is fairly strong for both car types: for midsize cars, $r = 0.80475$; for large cars, $r = 0.84261$. **(d)** For midsize cars: $\hat{y} = 7.28248 + 0.98263x$. For large cars: $\hat{y} = 2.21774 + 1.32258x$. The (dashed) regression line for large cars lies above the line for midsize cars (over the part of the graph where most of the points are found)—which agrees with the impression mentioned in (a). **(e)** $\hat{y} = 26.9$ mile per gallon for a midsize car; $\hat{y} = 28.7$ for a large car.

2.90 (a) $\hat{y} = 13.82 + 1.079x$; $\hat{y} = 504.8$ when $x = 455$. **(b)** Hawaii, with median SATV 404 and median SATM 481, is the outlier (circled). The predicted SATM for Hawaii is 449.7 (lower than actual); the residual is 31.3.

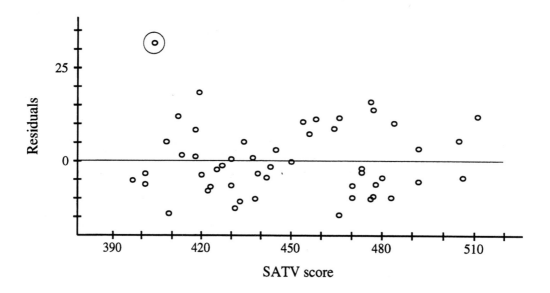

CHAPTER 3 SOLUTIONS

Section 3.1

3.1 The population is employed adult women, the sample is the 48 club members who returned the survey.

3.2 (a) An individual is a person; the population is all adult U.S. residents. **(b)** An individual is a household; the population is all U.S. households. **(c)** An individual is a voltage regulator; the population is all the regulators in the last shipment.

3.3 Only persons with an strong opinion on the subject—strong enough that they are willing to spend the time, and 50 cents—will respond to this advertisment.

3.4 Letters to legislators are an example of a voluntary response sample—the proportion of letters opposed to the insurance should not be assumed to be a fair representation of the attitudes of the congresswoman's constituents.

3.5 Starting with 01 and numbering down the columns, one chooses 04–Bonds, 10–Fleming, 17–Liao, 19–Naber, 12–Goel, and 13–Gomez.

3.6 Starting with 01 and numbering down the columns, one chooses 19–Laskowsky, 26–Rodriguez, 06–Castillo, and 09–Gonzalez.

3.7 Labeling from 001 to 440, we select 400, 077, 172, 417, 350, 131, 211, 273, 208, and 074.

3.8 Assign 01 to 30 to the students (in alphabetical order). The exact selection will depend on the starting line chosen in Table B; starting on line 123 gives 08–Ghosh, 15–Jones, 07–Fisher, and 27–Shaw. Assigning 0–9 to the faculty members gives (from line 109) 3–Fernandez and 6–Lightman. (We could also number facult from 01 to 10, but this requires looking up 2-digit numbers).

3.9 Label the 500 midsize accounts from 001 to 500, and the 4400 small accounts from 0001 to 4400. We first encounter numbers 417, 494, 322, 247, and 097 for the midsize group, then 3698, 1452, 2605, 2480, and 3716 for the small group.

3.10 (a) Households without telephones, or with unlisted numbers. Such households would likely be made up of poor individuals (who cannot afford a phone), those who choose not to have phones, and those who do not wish to have their phone number published. **(b)** Those with unlisted numbers would be included in the sampling frame when a random-digit dialer is used.

3.11 The higher no-answer was probably the second period—more families are likely to be gone for vacations, etc. Nonresponse of this type might underrepresent those who are more affluent (and are able to travel).

3.12 Form A would draw the higher negative response. It is phrased to produce a negative reaction: "giving huge sums of money" versus "contributing," and giving "to condidates" rather than "to campaigns." Also, form B presents both sides of the issue, allowing for special interest groups to have "a right to contribute."

3.13 The increased sample size gives more accurate information about the population (for example, a more accurate estimate of how many voters favor Candidate A over Candidate B).

3.14 The population is words in novels by Tom Wolfe; the sample is the first 250 words on the randomly selected page in the randomly selected novel. The variable is the length of a word.

3.15 **(a)** An individual is a small business; the specific population is "eating and drinking establishments" in the large city. **(b)** An individual is an adult; the Congressman's constituents are the *desired* population; the letter-writers are a voluntary sample and do not represent that population well. **(c)** Individual: auto insurance claim; the population is all the auto insurance claims filed in a given month.

3.16 Hite's questionnaires were distributed through women's groups, and thus could not include women who don't belong to such groups. Furthermore, since response was voluntary, the sample is probably biased toward those with strong feelings on the subjects. Thus, Hite's reported percentages are likely to be higher than the true percentages for the whole population of adult American women.

3.17 The call-in poll is faulty in part because it is a voluntary sample. Furthermore, even a small charge like 50 cents can discourage some people from calling in—especially poor people. Reagan's Republican policies appealed to upper-class voters, who would be less concerned about a 50 cent charge than lower-class voters who might favor Carter.

3.18 The interviewers would only get responses at households where someone was home during normal working hours. People in such households are more likely to have time to bake bread.

3.19 Number the bottles across the rows from 01 to 25, then select 12–B0986, 04–A1101, and 11–A2220. (Note: if numbering is done down columns instead, the sample will be A1117, B1102, and A1098.)

3.20 The blocks are already marked; select three-digit numbers and ignore those that do not appear on the map. This gives 214, 313, 409, 306, and 511.

3.21 (a) False—if it were true, then after looking at 39 digits, we would know whether or not the 40th digit was a 0, contrary to property 2. **(b)** True—there are 100 pairs of digits 00 through 99, and all are equally likely. **(c)** False—0000 is just as likely as any other string of four digits.

3.22 (a) Split the 200 addresses into 5 groups of 40 each. Looking for 2-digit numbers from 01 to 40, we find 35, and so take 35, 75, 115, 155, and 195. **(b)** Every address has a 1-in-40 chance of being selected, *but* not every subset has an equal chance of being picked—for example, 01, 02, 03, 04, and 05 cannot be selected by this method.

3.23 It is *not* an SRS. In order to be an SRS, every possible sample of 250 must have an equal chance of being chosen, and this is not the case—a group consisting of 250 female engineers will not be picked, for example.

3.24 (a) This question will likely elicit more responses against gun control (that is, more people will choose 2). The two options presented are too extreme; no middle position on gun control is allowed. **(b)** The phrasing of this question will tend to make people respond in favor of a nuclear freeze. Only one side of the issue is presented. **(c)** The wording is too technical for many people to understand—and for those that *do* understand it, it is slanted because it suggests reasons why one should support recycling. It could be rewritten to something like: "Do you support economic incentives to promote recycling?"

3.25 A smaller sample gives less information about the population. "Men" constituted only about one-third of our sample, so we know less about that group than we know about all adults.

Section 3.2

3.26 It is not an experiment: the political scientist is merely observing a characteristic (party preference) of a group of subjects. Sex is explanatory, and the party voted for in the last election is the response variable.

3.27 (a) The liners are the experimental units. **(b)** The heat applied to the liners is the factor; the levels are 250°F, 275°F, 300°F, and 325°F. **(c)** The force required to open the package is the response variable.

3.28 (a) This is an experiment, since the teacher imposes treatments (instruction method). **(b)** The explanatory variable is the method used (computer software or standard curriculum), and the response is the change in reading ability.

3.29 (a) The experimental units are the batches of the product; the yield of each batch is the response variable. **(b)** There are two factors: temperature (with 2 levels) and stirring rates (with 3 levels), for a total of 6 treatments. **(c)** Since two experimental units will be used for each treatment, we need 12.

		Factor B: Stirring Rates		
		60 rpm	90 rpm	120 rpm
Factor A:	50°C	1	2	3
Temperature	60°C	4	5	6

3.30 (a) In a serious case, when the patient has little chance of surviving, a doctor might choose not to recommend surgery; it might be seen as an unnecessary measure, bringing expense and a hospital stay with little benefit to the patient.
(b)

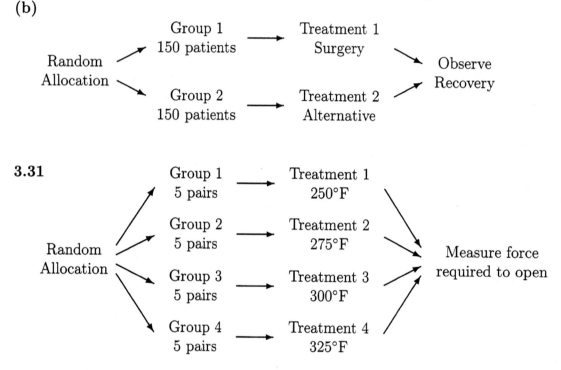

3.32 (a) Randomly select 20 women for Group 1, which will see the "childcare" version of Company B's brochure, and assign the other 20 women to Group 2 (the "no childcare" group). Allow all women to examine the appropriate brochures, and observe which company they choose. Compare the number from Group 1 who choose Company B with the corresponding number from Group 2. **(b)** Numbering from 01 to 40, Group 1 is 05–Cansico, 32–Roberts, 19–Hwang, 04–Brown, 25–Lippman, 29–Ng, 20–Iselin, 16–Gupta, 37–Turing, 39–Williams, 31–Rivera, 18–Howard, 07–Cortez, 13–Garcia, 33–Rosen, 02–Adamson, 36–Travers, 23–Kim, 27–McNeill, and 35–Thompson.

3.33 Number the liners from 01 to 20, then take Group 1 to be 16, 04, 19, 07, and 10; Group 2 is 13, 15, 05, 09, and 08; Group 3 is 18, 03, 01, 06, and 11. The others are in Group 4.

3.34 Randomly assign 6 students to each of Groups 1, 2, 3, 4, 5 and 6 (place the first 6 selected students in Group 1, the next 6 in Group 2, and so on). Each group will watch the version of the television show with the corresponding treatment (as numbered in Figure 3.2). Then observe their responses to the questions about their attitude toward the product, etc.

Group 1 is students 05, 16, 17, 20, 16, and 32; Group 2 is 04, 25, 29, 31, 18, and 07; Group 3 is 13, 33, 02, 36, 23, and 27; Group 4 is 35, 21, 26, 08, 10, and 11; Group 5 is 15, 12, 14, 09, 24, and 22; the rest are in Group 6.

3.35 If this year is considerably different in some way from last year, we cannot compare electricity consumption over the two years. For example, if this summer is warmer, the customers may run their air conditioners more often. The possible differences between the two years would confound the effects of the treatments.

3.36 The second design is an experiment—a treatment is imposed on the subjects. The first is a study; it may be confounded by the types of men in each group. In spite of the researcher's attempt to match "similar" men from each group, those in the first group (who exercise) could be somehow different from men in the non-exercising group.

3.37 There almost certainly was *some* difference between the sexes and between blacks and whites; the difference between men and women was so large that it is unlikely to be due to chance. For black and white students, however, the difference was small enough that it could be attributed to random variation.

3.38 Because the experimenter knew which subjects had learned the meditation techniques, he (or she) may have had some expectations about the outcome of the experiment: if the experimenter believed that meditation was beneficial, he may subconsciously rate that group as being less anxious.

3.39 (a) If only the new drug is administered, and the subjects are then interviewed, their responses will not be useful, because there will be nothing to compare them to: How much "pain relief" does one expect to experience? **(b)** Randomly assign 20 patients to each of three groups: Group 1, the placebo group; Group 2, the aspirin group; and Group 3, which will receive the new medication. After treating the patients, ask them how much pain relief they feel, and then compare the average pain relief experienced by each group. **(c)** The subjects should certainly not know what drug they are getting—a patient told that she is receiving a placebo, for example, will probably not expect any pain relief. **(d)** Yes—presumably, the researchers would like to conclude that the new medication is better than aspirin. If it is not double-blind, the interviewers may subtly influence the subjects into giving responses that support that conclusion.

3.40 (a) Ordered by increasing weight, the five blocks are (1) Williams–22, Deng–24, Hernandez–25, and Moses–25; (2) Santiago–27, Kendall–28, Mann–28, and Smith–29; (3) Brunk–30, Obrach–30, Rodriguez–30, and Loren–32; (4) Jackson–33, Stall–33, Brown–34, and Cruz–34; (5) Birnbaum–35, Tran–35, Nevesky–39, and Wilansky–42. **(b)** The exact randomization will vary with the starting line in Table B. Different methods are possible; perhaps the simplest is to number from 1 to 4 within each block, then assign the members of block 1 to a weight-loss treatment, then assign block 2, etc. For example, starting on line 133, we assign 4–Moses to treatment A, 1–Williams to B, and 3–Hernandez to C (so that 2–Deng gets treatment D), then carry on for block 2, etc. (either continuing on the same line, or starting over somewhere else).

3.41 (a)

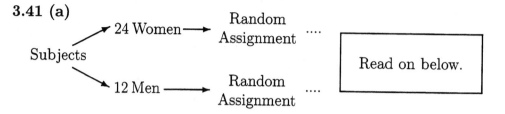

Following the "random assignment," the top branch of the diagram splits into six groups of 4 women each, and the bottom splits into six groups with 2 men in each. Each group receives the appropriate treatment (1–6), and then the subjects' attitudes about the product, etc., are measured. **(b)** Number the women from 01 to 24, and the men from 01 to 12. First we look for 20 women's numbers, and find: 12, 13, 04, 18, 19, 24, 23, 16, 02, 08, 17, 21, 10, 05, 09, 06, 01, 20, 22, and 07. Continuing on to find 10 men's numbers, we get 05, 09, 07, 02, 01, 08, 11, 06, 12, and 04. Then, for example, Women's Group 1 will be the first four women selected, namely 12, 13, 04, and 18. This gives us the following layout:

Women

01	5	07	5	13	1	19	2
02	3	08	3	14	6	20	5
03	6	09	4	15	6	21	3
04	1	10	4	16	2	22	5
05	4	11	6	17	3	23	2
06	4	12	1	18	1	24	2

Men

01	3	07	2
02	2	08	3
03	6	09	1
04	5	10	6
05	1	11	4
06	4	12	5

3.42 For each person, randomly decide which hand they should use first—either by flipping a coin (heads: right hand first, tails: left hand first) or by taking digits from Table B (even: right, odd: left).

3.43 The randomization will vary with the starting line in Table B. *Completely randomized design:* Randomly assign 10 students to "Group 1" (which has the trend-highlighting software) and the other 10 to "Group 2" (which does not). Compare the performance of Group 1 with that of Group 2. *Matched pairs design:* Each student does the activity twice, once with the software and once without. Randomly decide (for each student) whether they have the software the first or second time. Compare performance with the software and without it. (This randomization can be done by flipping a coin 20 times, or by picking 20 digits from Table B, and using the software first if the digit is even, etc.) *Alternate Matched pairs design:* Again, all students do the activity twice. Randomly assign 10 students to Group 1 and 10 to Group 2. Group 1 uses the software the first time; Group 2 uses the software the second time.

3.44 (a) Assign 10 subjects to Group 1 (the 70° group) and the other 10 to Group 2 (which will performs the task in the 90° condition). Record the number of correct insertions in each group. **(b)** All subjects will perform the task twice—once in each temperature condition. Randomly choose which temperature each subject works in first, either by flipping a coin, or by placing 10 subjects in Group 1 (70°, then 90°) and the other 10 in Group 2.

3.45 (a) Each subject takes both tests; the order in which the tests are taken is randomly chosen. **(b)** Take 22 digits from Table B. If the first digit is even, subject 1 takes the BI first; if it is odd, he or she takes the ARSMA first. (Or, administer the BI first if the first digit is 0–4, and the ARSMA first if it is 5–9).

3.46 (a)

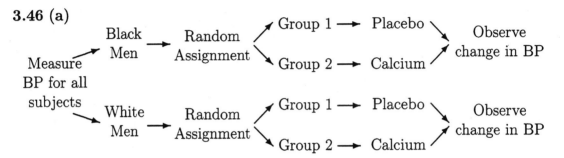

(b) A larger group gives more information—when more subjects are involved, the random differences between individuals have less influence, and we can expect the average of our sample to be a better representation of the whole population.

3.47 (a) It is an experiment (albeit a poorly designed one), because a treatment (herbal tea) is imposed on the subjects. **(b)** No, it is a study—the scores on the English test are merely observed for the various subjects.

3.48 (a) The subjects are the 210 children. **(b)** The factor is the "choice set"; there are three levels (2 milk/2 fruit drink, 4 milk/2 fruit drink, and 2 milk/4 fruit drink). **(c)** The response variable is the choice made by each child.

3.49 (a) The subjects are the physicians, the factor is medication (with two levels—aspirin and placebo), and the response is observing health, specifically whether the subjects have heart attacks or not.
(b)

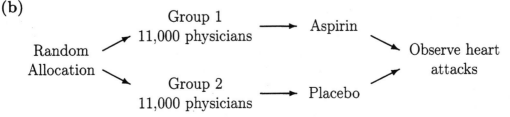

3.50

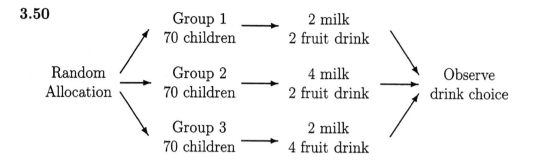

3.51 (a) Randomly assign 20 men to each of two groups. Record each subject's blood pressure, then apply the treatments: a calcium supplement for Group 1, and a placebo for Group 2. After sufficient time has passed, measure blood pressure again and observe any change. **(b)** Number from 01 to 40 down the columns. Group 1 is 18–Howard, 20–Imrani, 26–Maldonado, 35–Tompkins, 39–Willis, 16–Guillen, 04–Bikalis, 21–James, 19–Hruska, 37–Tullock, 29–O'Brian, 07–Cranston, 34–Solomon, 22–Kaplan, 10–Durr, 25–Liang, 13–Fratianna, 38–Underwood, 15–Green, and 05–Chen.

3.52 Label the children from 001 to 210, then consider three digits at a time. The first five children in Group 1 are numbers 119, 033, 199, 192, and 148.

3.53 Responding to a placebo does not imply that the complaint was not "real"—38% of the placebo group in the gastric freezing experiment improved, and those patients really had ulcers. The placebo effect is a *psychological* response, but it may make an actual *physical* improvement in the patient's health.

Chapter Review

3.54 (a) The population is Ontario residents; the sample is the 61,239 people interviewed. **(b)** The sample size is very large, so if there were large numbers of both sexes in the sample—this is a safe assumption since we are told this is a "random sample"—these two numbers should be fairly accurate reflections of the values for the whole population.

3.55 (a) Explanatory variable: treatment method; response: survival times. **(b)** No treatment is actively imposed; the women (or their doctors) chose which treatment to use. **(c)** Doctors may make the decision of which treatment to recommend based in part on how advanced the case is. Some might be more likely to recommend the older treatment for advanced cases, in which case the chance of recovery is lower. Other doctors might view the older treatment as not being worth the effort, and recommend the newer method as a way of providing *some* hope for recovery while minimizing the trauma and expense of major surgery.

3.56 No, it is a study—no treatment is imposed; the researchers simply measure the fitness of the executives.

3.57 (a) Label the students from 0001 to 3478. **(b)** Taking four digits at a time gives 2940, 0769, 1481, 2975, and 1315.

3.58 A stratified random sample would be useful here; one could select 50 faculty members from each level. Alternatively, select 25 (or 50) institutions of each size, then choose 2 (or 1) faculty members at each institution.

If a large proportion of faculty in your state work at a particular class of institution, it may be useful to stratify unevenly. If, for example, about 50% teach at Class I institutions, you may want half your sample to come from Class I institutions.

3.59 (a) One possible population: all full-time undergraduate students in the fall term on a list provided by the Registrar. **(b)** A stratified sample with 125 students from each year is one possibility. **(c)** Mailed questionnaires might have high non-response rates. Telephone interviews exclude those without phones, and may mean repeated calling for those that are not home. Face-to-face interviews might be more costly than your funding will allow.

3.60 (a) The chicks are the experimental units; weight gain is the response variable. **(b)** There are two factors: corn variety (2 levels) and percent of protein (3 levels). This makes 6 treatments, so 60 chicks are required.

		Factor B: Protein Level	
	12%	16%	20%
Factor A: opaque-2	1	2	3
Corn Variety floury-2	4	5	6

Weigh chicks; Random Allocation

Group 1 / 10 chicks → opaque-2 / 12% protein

Group 2 / 10 chicks → opaque-2 / 16% protein

Group 3 / 10 chicks → opaque-2 / 20% protein

Group 4 / 10 chicks → floury-2 / 12% protein

Group 5 / 10 chicks → floury-2 / 16% protein

Group 6 / 10 chicks → floury-2 / 20% protein

Record weight change

3.61 (a)

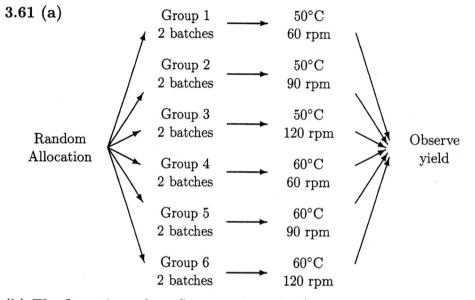

Group 1 — 2 batches → 50°C 60 rpm
Group 2 — 2 batches → 50°C 90 rpm
Group 3 — 2 batches → 50°C 120 rpm
Group 4 — 2 batches → 60°C 60 rpm
Group 5 — 2 batches → 60°C 90 rpm
Group 6 — 2 batches → 60°C 120 rpm

Random Allocation → Observe yield

(b) The first 10 numbers (between 01 and 12) are 06, 09, 03, 05, 04, 07, 02, 08, 10, and 11. So the 6th and 9th batches will receive treatment 1; batches 3 and 5 will be processed with treatment 2, etc.

3.62 The factors are whether or not the letter has a ZIP code (2 levels: yes or no), and the time of day the letter is mailed. The number of levels for the second factor may vary.

   To deal with lurking variables, all letters should be the same size and should be sent to the same city, and the day on which a letter is sent should be randomly selected. Because most post offices have shorter hours on Saturdays, one may wish to give that day some sort of "special treatment" (it might even be a good idea to have the day of the week be a *third* factor in this experiment).

3.63 Each subject should taste both kinds of cheeseburger, in a randomly selected order, and then be asked about their preference. Both burgers should have the same "fixings" (ketchup, mustard, etc.). Since some subjects might be able to identify the cheeseburgers by appearance, one might need to take additional steps (such as blindfolding, or serving only the center part of the burger) in order to make this a true "blind" experiment.

3.65 It means that the correlation is large enough (presumably, though not necessarily, in the positive direction) that it is unlikely to have occurred just by chance.

CHAPTER 4 SOLUTIONS

Section 4.1

4.1 7.2% is a statistic.

4.2 2.5003 is a parameter; 2.5009 is a statistic.

4.3 48% is a statistic; 52% is a parameter.

4.4 Both 335 and 289 are statistics.

4.5 The appearance of the histogram will vary from experiment to experiment. For comparison, here is the sampling distribution (assuming *p* really is 0.5). Answers for (b) will probably not resemble this much, but for (c), they may be fairly close.

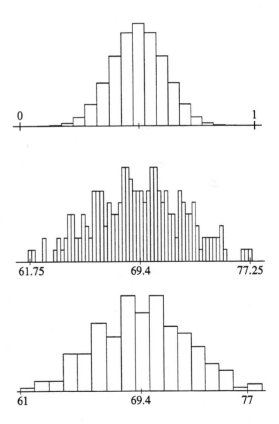

4.6 (a) The scores will vary depending on the starting row. Note that the smallest possible mean is 61.75 (from the sample 58, 62, 62, 65) and the largest is 77.25 (from 73, 74, 80, 82). Answers to **(b)** and **(c)** will vary; shown are two views of the sampling distribution. The first shows all possible values of the experiment (so the first rectangle is for 61.75, the next is for 62.00, etc.); the other shows values grouped from 61 to 61.75, 62 to 62.75, etc. (which makes the histogram less bumpy). The tallest rectangle in the first picture is 8 units; in the second, the tallest is 28 units.

4.7 (a) Table is below; histogram not shown. **(b)** The histogram actually does *not* appear to have a normal shape. The sampling distribution is quite normal in appearance, but even a sample of size 100 does not *necessarily* show it. **(c)** The mean of $\hat{p}$ is 0.10194. The bias seems to be small. **(d)** The mean of the sampling distribution should be $p = 0.10$. **(e)** The mean would still be 0.10, but the spread would be smaller.

p	$\hat{p}$	Count	p	$\hat{p}$	Count	p	$\hat{p}$	Count
9	0.045	1	18	0.090	12	24	0.120	10
13	0.065	3	19	0.095	9	25	0.125	4
14	0.070	2	20	0.100	7	26	0.130	1
15	0.075	5	21	0.105	5	27	0.135	2
16	0.080	11	22	0.110	6	28	0.140	2
17	0.085	12	23	0.115	7	30	0.150	1

4.8 (a) Large bias and large variability. **(b)** Small bias and small variability. **(c)** Small bias, large variability. **(d)** Large bias, small variability.

4.9 (a) Since the smallest number of total tax returns (i.e., the smallest population) is still more than 100 times the sample size, the variability will be the (approximately) the same for all states. **(b)** Yes, it will change—the sample taken from Wyoming will be about the same size, but the sample in, e.g., California will be considerably larger, and therefore the variability will decrease.

4.11 There are *21* 0s among the first 200 (to be specific, $3+5+6+4+3$ respectively in the first five rows), so $\hat{p} = 0.105$.

4.12 (a) 0. **(b)** 1. **(c)** 0.01. **(d)** 0.6 (or 0.99).

4.13 (a) 0.1. **(b)** 0.5.

4.14 (a) 0.633. **(b)** 0.079.

4.15 (a) Use digits 0 and 1 (or any other 2 of the 10 digits) to represent the presence of egg masses. Reading the first 10 digits from line 116, for example, gives YNNNN NNYNN— 2 square yards with egg masses, 8 without—so $\hat{p} = 0.2$. **(b)** The stemplot *might* look like the one on the right (which is close to the sampling distribution of $\hat{p}$). **(c)** The mean would be $p = 0.2$. **(d)** 0.4.

```
0.0 | 00
0.0 | 55555
0.1 | 000000
0.1 | 5555
0.2 | 00
0.2 | 5
```

4.16 (a) Use 0–3 to represent persons who would answer "yes." Looking at the first 20 digits on line 136 gives YNNYY NNNNY NNYNN YNNNN—6 yes and 14 no, so $\hat{p} = 0.3$. **(b)** Most answers should fall between 0.3 and 0.5. **(c)** 0.4. **(d)** 0.5.

4.17 Assuming that the poll's sample size was less than 780,000—10% of the population of New Jersey—the variability would be practically the same for either population. (The sample size for this poll would have been considerably less than 780,000.)

4.19 (a) Most answers should fall between 0.3 and 0.7. **(b)** The *exact* probability is $\frac{2}{3}$; most answers should fall between 0.47 and 0.87.

4.20 Due to either: 0.66. Due to something else: 0.34.

4.21 (a) 0.65. **(b)** 0.38. **(c)** 0.62.

4.22 (a) The sum is 1, as it should be since all possible responses (top 20%, etc.) are listed. **(b)** 0.59. **(c)** 0.64.

4.23 $18/38 = 9/19$, or about 0.4737.

Section 4.2

4.24 (a) 1%. **(b)** All probabilities are between 0 and 1, and they add up to 1. **(c)** 0.94. **(d)** 0.86. **(e)** "$X \geq 4$" or "$X > 3$" or "$X = 4$ or $X = 5$." The probability is 0.06.

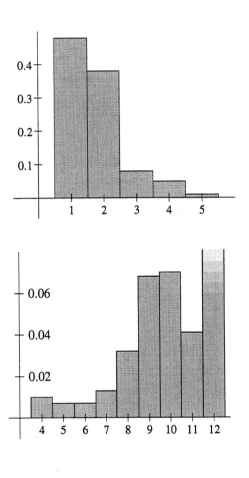

4.25 (a) 75.2%. **(b)** All probabilities are between 0 and 1, and they add up to 1. The histogram does not show the whole rectangle for year 12, so that the details of years 4–11 can be seen more clearly. **(c)** 0.983. **(d)** 0.976. **(e)** "$X \geq 9$" or "$X > 8$." The probability is 0.931.

4.26 $2/6 = 1/3$, or about 0.3333.

4.27 $5/26$, or about 0.1923.

4.28 (a) GGG, GGB, GBG, BGG, BBG, BGB, GBB, BBB; each arrangement has (approximate) probability $1/8 = 0.125$. **(b)** Three of the eight arrangements have two (and only two) girls, so $P(X = 2) = 3/8 = 0.375$. **(c)** On right.

Value of X	0	1	2	3
Probability	$\frac{1}{8}$	$\frac{3}{8}$	$\frac{3}{8}$	$\frac{1}{8}$

4.29 (a) 10,000. **(b)** $1/10000 = 0.0001$. **(c)** There are 10 (0000, 1111, 2222, ..., 9999); the probability that you get one is $10/10000 = 1/1000 = 0.001$.

4.30 (a) Below. **(b)** The mean is $\mu = 7$—the obvious balancing point of the histogram.

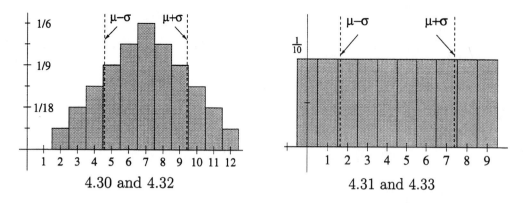

4.30 and 4.32   4.31 and 4.33

4.31 $\mu = 4.5$—the balancing point for the probability histogram (halfway between 4 and 5).

4.32 $\sigma^2 = 35/6$ (about 5.8333), so σ is about 2.415. See the histogram above. The probability of getting an outcome within σ of μ is the probability of rolling a 5, 6, 7, 8 or 9—a total of 29/36, or about 0.8056.

4.33 $\sigma = \sqrt{8.25} = 2.8723$. See the histogram above. The numbers within one standard deviation of μ are 2, 3, 4, 5, 6, and 7; the probability is 0.60.

4.34 (a) The curve forms a 1×1 square; the area is 1. **(b)** $P = 0.25$. **(c)** $P = 0.8$. **(d)** $P = 0.2$.

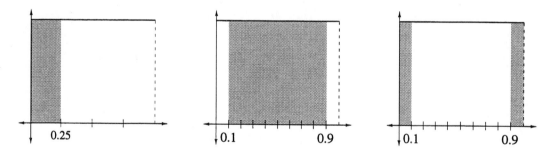

4.35 (a) The area of a triangle is $\frac{1}{2}bh = \frac{1}{2}(2)(1) = 1$. **(b)** $P = 0.5$. **(c)** $P = 0.125$.

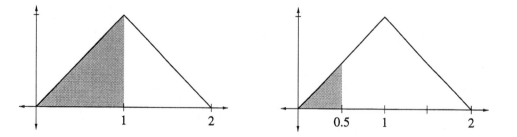

4.36 (a) $P(\hat{p} \geq 0.5) = P(Z \geq \frac{0.5-0.3}{0.023}) = P(Z \geq 8.7) = 0$. **(b)** $P(\hat{p} < 0.25) = P(Z < -2.17) = 0.0149$. **(c)** $P(0.25 \leq \hat{p} \leq 0.35) = P(-2.17 \leq Z \leq 2.17) = 0.9703$.

4.37 $\mu = 2.25$—lower than a B. $\sigma = \sqrt{1.3875} = 1.1779$.

4.38 (a) $\mu = -\$0.25$ (player loses 25 cents). **(b)** The casino makes 25 cents for every dollar bet (in the long run). **(c)** $\sigma = \sqrt{1.6875} = 1.299$.

4.39 $\mu = 11.251$, $\sigma = \sqrt{2.444} = 1.563$.

4.40 (a) The height should be $\frac{1}{2}$, since the area under the curve must be 1. The density curve is on the right. **(b)** $\mu = 1$. **(c)** $P(X \leq 1) = \frac{1}{2}$. **(d)** $P(0.5 < X < 1.3) = 0.4$. **(e)** $P(X \geq 0.8) = 0.6$.

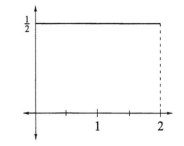

4.41 (a) $P(\hat{p} \geq 0.16) = P(Z \geq 1.087) = 0.1385$. **(b)** $P(0.14 \leq \hat{p} \leq 0.16) = P(1.087 \leq Z \leq 1.087) = 0.7230$.

Section 4.3

4.42 (a) $\mu = p = 0.15$, $\sigma = \sqrt{(0.15)(0.85) \div 1540} = 0.0091$. **(b)** The population (US adults) is considerably larger than 10 times the sample size (1540). **(c)** $np = 231$, $n(1 - p) = 1309$—both are much bigger than 10. **(d)** $P(0.13 < \hat{p} < 0.17) = P(-2.198 < Z < 2.198) = 0.9722$. **(e)** To achieve $\sigma = 0.0045$, we need a sample four times as large: 6160.

4.43 (a) $\mu = p = 0.4$, $\sigma = \sqrt{(0.4)(0.6) \div 1785} = 0.0116$. **(b)** The population (US adults) is considerably larger than 10 times the sample size. **(c)** $np = 714$, $n(1-p) = 1071$—both are much bigger than 10. **(d)** $P(0.37 < \hat{p} < 0.43) = P(-2.586 < Z < 2.586) = 0.9904$. Over 99% of all samples should give $\hat{p}$ within $\pm 3\%$ of the true population proportion.

4.44 For $n = 200$: $\sigma = 0.02525$, and the probability is $P = 0.5704$. For $n = 800$: $\sigma = 0.01262$ and $P = 0.8858$. For $n = 3200$: $\sigma = 0.0631$ and $P = 0.9984$. Larger sample sizes give more accurate results (the sample proportions are more likely to be close to the true proportion).

4.45 For $n = 300$: $\sigma = 0.02828$ and $P = 0.7108$. For $n = 1200$: $\sigma = 0.01414$ and $P = 0.9660$. For $n = 4800$: $\sigma = 0.00707$ and $P = 1$ (approximately). Larger sample sizes give more accurate results (the sample proportions are more likely to be close to the true proportion).

4.46 (a) $\mu = 0.52$, $\sigma = 0.02234$. **(b)** np and $n(1 - p)$ are 260 and 240 respectively. $P(\hat{p} \geq 0.50) = P(Z \geq -0.8951) = 0.8159$.

4.47 (a) 0.86 (86%). **(b)** We use the normal approximation (Rule of Thumb 2 is *just* satisfied—$n(1 - p) = 10$). The standard deviation is 0.03, and $P(\hat{p} \leq 0.86) \doteq P(Z \leq -1.33) = 0.0918$. (NOTE: The exact probability is 0.1239.) **(c)** Even when the claim is correct, there will be some variation in sample proportions. In particular, in about 10% of samples we can expect to observe 86 or fewer orders shipped on time.

4.48 Comparing the results of "Rule of Thumb 2," we see that it is clearly satisfied in the telephone number problem, and just barely satisfied in the mail-order problem—so the approximation is more accurate in the first of these.

4.49 (a) $np = 1190$. (b) $P(< 1200 \text{ students accept}) = P(\hat{p} < \frac{1200}{1700}) \doteq P(Z < 0.5293) = 0.7019$. (c) $P(\frac{1150}{1700} < \hat{p} < \frac{1250}{1700}) \doteq P(-2.117 < Z < 3.176) = 0.9847$.

4.50 (a) $np = 100$. (b) $P(\hat{p} \geq \frac{90}{500}) \doteq P(Z \geq -1.118) = 0.8686$.

4.51 (a) $np = 66$ and $n(1-p) = 234$, so Rule of Thumb 2 is satisfied. $P(\hat{p} > 0.20) = P(Z > -0.8362) = 0.7985$. (b) $P(\hat{p} > 0.30) = P(Z > 3.345) = 0.0004$.

4.52 (a) $P(\hat{p} \leq 0.70) \doteq P(Z \leq -1.155) = 0.1241$. (b) $P(\hat{p} \leq 0.70) \doteq P(Z \leq -1.826) = 0.0339$. (c) The test must contain 400 questions. (d) The answer is the same for Laura.

4.53 (a) $np = 80$. (b) Still assuming that $p = 0.04$, $P(\hat{p} \geq \frac{75}{2000}) = P(Z \geq -0.57) = 0.7157$.

4.54 (a) $np = (15)(0.3) = 4.5$—this fails Rule of Thumb 2. (b) The population size (316) is not at least 10 times as large as the sample size (50)—this fails Rule of Thumb 1.

Section 4.4

4.55 It may be binomial if we assume that there are no twins or other multiple births among the next 20 (this would violate requirement 2—independence—of the binomial setting), and that for all births, the probability that the baby is female is the same (requirement 4).

4.56 No—the number of observations is not fixed.

4.57 No—since she receives instruction after incorrect answers, her probability of success is likely to increase.

4.58 Assuming that Joe's chance of winning the lottery is the same every week, and that a year consists of 52 weeks (observations), this would be binomial.

4.59 (a) 0, 1, ..., 5.
(b) $P(X = 0) = \binom{5}{0}(.25)^0(.75)^5 = 0.23730$;
$\quad P(X = 1) = 0.39551$;
$\quad P(X = 2) = 0.26367$;
$\quad P(X = 3) = 0.08789$;
$\quad P(X = 4) = 0.01465$;
$\quad P(X = 5) = 0.00098$.

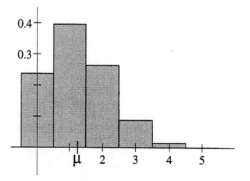

4.60 (a) $\binom{15}{3}(0.3)^3(0.7)^{12} = 0.17004$. **(b)** $\binom{15}{0}(0.3)^0(0.7)^{15} + \cdots + \binom{15}{3}(0.3)^3(0.7)^{12} = 0.29687$.

4.61 $P(X = 10) = \binom{20}{10}(0.8)^{10}(0.2)^{10} = 0.00203$.

4.62 (a) $n = 10$ and $p = 0.25$. **(b)** $\binom{10}{2}(0.25)^2(0.75)^8 = 0.28157$. **(c)** $P(X \le 2) = \binom{10}{0}(0.25)^0(0.75)^{10} + \cdots + \binom{10}{2}(0.25)^2(0.75)^8 = 0.52559$.

4.63 $\mu = \frac{5}{4} = 1.25$, $\sigma = \sqrt{\frac{15}{16}} = 0.96825$. See exercise 4.59 for plot.

4.64 (a) $\mu = 4.5$. **(b)** $\sigma = \sqrt{3.15} = 1.77482$. **(c)** If $p = 0.1$, then $\sigma = \sqrt{1.35} = 1.16190$. If $p = 0.01$, then $\sigma = \sqrt{0.1485} = 0.38536$. As p gets close to 0, σ gets closer to 0.

4.65 (a) $\mu = 16$ (if $p = 0.8$). **(b)** $\sigma = \sqrt{3.2} = 1.78885$. **(c)** If $p = 0.9$, then $\sigma = \sqrt{1.8} = 1.34164$. If $p = 0.99$, then $\sigma = \sqrt{0.198} = 0.44497$. As p gets close to 1, σ gets closer to 0.

4.66 $\mu = 2.5$, $\sigma = \sqrt{1.875} = 1.36931$.

4.67 (a) $n = 20$ and $p = 0.25$. **(b)** $\mu = 5$. **(c)** $\binom{20}{5}(0.25)^5(0.75)^{15} = 0.20233$.

4.68 (a) $n = 6$ and $p = 0.65$. **(b)** X takes values from 0 to 6. **(c)** $P(X = 0) = 0.00184$, $P(X = 1) = 0.02048$, $P(X = 2) = 0.09510$, $P(X = 3) = 0.23549$, $P(X = 4) = 0.32801$, $P(X = 5) = 0.24366$, $P(X = 6) = 0.07542$. **(d)** $\mu = 3.9$, $\sigma = \sqrt{1.365} = 1.16833$.

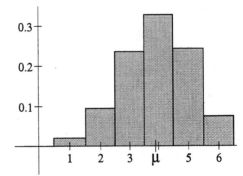

4.69 (a) The probability that all are assessed as truthful is $\binom{12}{0}(0.2)^0(0.8)^{12} = 0.06872$; the probability that at least one is reported to be a liar is $1 - 0.06872 = 0.93128$. **(b)** $\mu = 2.4$, $\sigma = \sqrt{1.92} = 1.38564$.

4.70 (a) X has a binomial distribution with $n = 20$ and $p = 0.99$. **(b)** $P(X = 20) = 0.81791$; $P(X < 20) = 0.18209$. **(c)** $\mu = 19.8$, $\sigma = \sqrt{0.198} = 0.44497$.

4.71 (a) $P(\text{card is black}) = P(\text{card is red}) = \frac{26}{52} = 0.5$. **(b)** There are 51 cards left, of which 25 are black. Now $P(\text{card is black}) = \frac{25}{51} = 0.49020$, and $P(\text{card is red}) = \frac{26}{51} = 0.50980$. **(c)** There are still 51 cards left; now 26 of those are black. This time $P(\text{card is black}) = \frac{26}{51} = 0.50980$, and $P(\text{card is red}) = \frac{25}{51} = 0.49020$.

4.72 (a) $P(\text{switch is bad}) = \frac{1000}{10000} = 0.1$; $P(\text{switch is OK}) = \frac{9000}{10000} = 0.9$. **(b)** 9999 switches remain, of which 999 are bad. Under these conditions, $P(\text{switch is bad}) = \frac{999}{9999} = 0.09990999...$, and $P(\text{switch is OK}) = \frac{9000}{9999} = 0.90009000....$ **(c)** Again,

9999 switches remain; this time 1000 are bad, so that $P(\text{switch is bad}) = \frac{1000}{9999} = 0.10001000...$, and $P(\text{switch is OK}) = \frac{8999}{9999} = 0.89998999....$

Section 4.5

4.73 The mean for 4.17(b) is the population mean from 4.17(a), namely -3.5%. The standard deviation is $\sigma/\sqrt{n} = 26\%/\sqrt{5} = 11.628\%$.

4.74 Mean: 18.6, standard deviation: $5.9/\sqrt{76} = 0.67678$. The normality of individual scores is not necessary for this to be true.

4.75 Mean: 40.125, standard deviation: 0.001; normality is not needed.

4.76 Standard deviation: 0.04619.

4.77 **(a)** $P(X \geq 21) = P(Z \geq \frac{21-18.6}{5.9}) = P(Z \geq 0.4068) = 0.3421.$ **(b)** $P(\overline{x} \geq 21) = P(Z \geq \frac{21-18.6}{5.9/\sqrt{50}}) = P(Z \geq 2.8764) = 0.0020.$

4.78 **(a)** $P(X < 295) = P(Z < -1) = 0.8413.$ **(b)** $P(\overline{x} < 295) = P(Z < -2.4495) = 0.0072.$

4.79 **(a)** $N(123, 0.04619).$ **(b)** $P(Z > 21.65)$—essentially 0.

4.80 $\overline{x}$ has approximately a $N(1.6, 0.0849)$ distibution; the probability is $P(Z > 4.71)$—essentially 0.

4.81 $\overline{x}$ (the mean return) has approximately a $N(9\%, 4.174\%)$ distibution; $P(\overline{x} > 15\%) = P(Z > 1.437) = 0.9247;$ $P(\overline{x} < 5\%) = P(Z < -0.9583) = 0.1690.$

4.82 The mean μ of the company's "winnings" (premiums) and their "losses" (insurance claims) is positive. Even though the company will lose a large amount of money on a small number of policyholders who die, it will gain a small amount on the majority. The law of large numbers guarantees that the average "winnings" minus "losses" will be close to μ, and overall the company will almost certainly show a profit.

4.83 **(a)** $N(55000, 4500/\sqrt{8}) = N(55000, 1591).$ **(b)** $P(Z < -2.011) = 0.0222.$

4.84 **(a)** $N(2.2, 0.1941).$ **(b)** $P(Z < -1.0304) = 0.1515.$ **(c)** $P(\overline{x} < \frac{100}{52} = P(Z < -1.4267) = 0.0768.$

4.85 **(a)** $P(Z < -1.5) = 0.0668.$ **(b)** $P(Z < -3) = 0.0013.$

4.86 $\mu + 2.33\sigma/\sqrt{n} = 1.4625.$

4.87 $\mu - 1.645\sigma/\sqrt{n} = 12.513.$

Section 4.6

4.88 The center line is at $\mu = 75°$; the control limits should be at $75° \pm 3\frac{\sigma}{\sqrt{4}}$, which means $74.25°$ and $75.75°$.

4.89 Center: 0.8750; control limits: 0.8750 ± 0.0016, i.e., 0.8734 and 0.8766.

4.90 (a) Center: 11.5; control limits: 11.2 and 11.8. **(b)** Output from Minitab (slightly modified) is below. Points outside control limits are circled; the ninth point of a run-of-9 is marked with a square. **(c)** Set B is from the in-control process; the process mean shifted suddenly for Set A, and it gradually drifted for Set C.

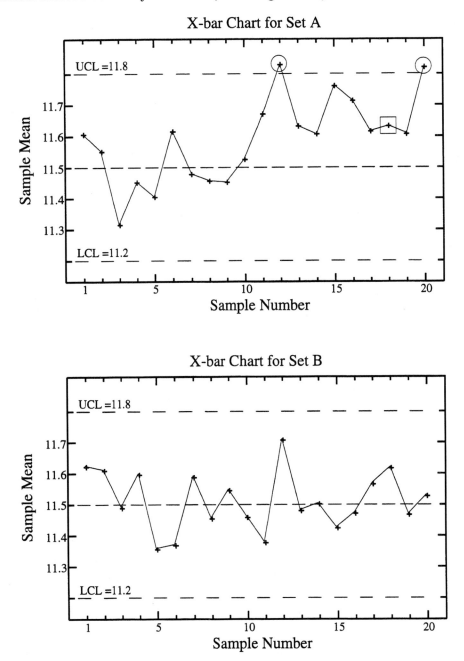

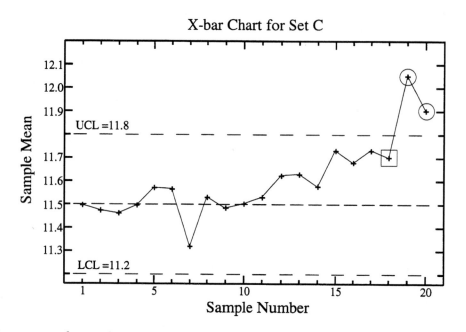

4.91 The mean of sample number 7 fell below the lower control limit (2.2037)—that would have been the time to correct the process.

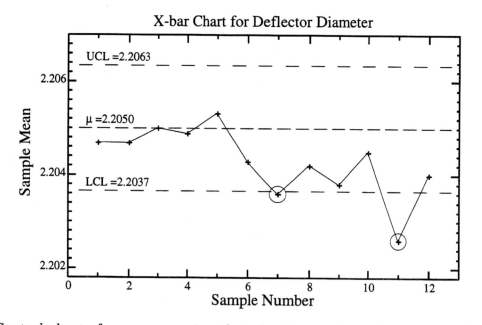

4.92 Control charts focus on ensuring that the *process* is consistent, not that the *product* is good. An in-control process may consistently produce some percentage of low-quality products. Keeping a process in control allows one to detect shifts in the distribution of the output (which may have been caused by some correctable error); it does not help in fixing problems that are inherent to the process.

4.93 (a) Center: 10; control limits: 7.922 and 12.078. **(b)** There are no runs that should concern us here—we would only be concerned with a run of samples with mean *less than* 10. Lot 13 signals that the process is out of control. The two samples that follow the bad one are fine, so it may be that whatever caused the low average for the 13th sample was an isolated incident (temperature fluctuations in the oven during the baking of that batch, or a bad batch of ingredients, perhaps). The operator should investigate to see if there is such an explanation, and try to remedy the situation if necessary.

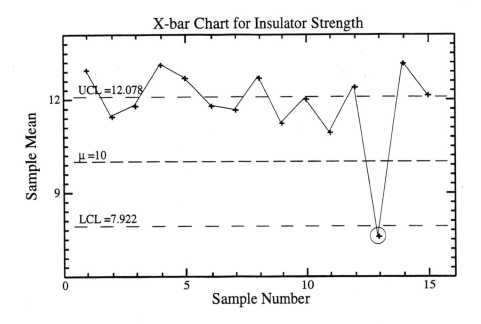

4.94 Center: 10; control limits: 10 ± 1.470, i.e., 8.530 and 11.470.

4.95 (a) About 43%: $P(\text{width} < 2.8 \text{ or width} > 3.2) = P(Z < -0.1913 \text{ or } Z > 2.4472) = 0.4241 + 0.0072 = 0.4313$. **(b)** Center: 3; control limits: 2.797 to 3.203.

4.96 The $\overline{x}$ chart only looks at the values of a sample mean, which we can expect to be less variable than individual observations. We use $\overline{x}$ charts to ensure that the *process mean* is close to the target; that does not imply that every screen will be close to the target.

99.7% of all individual tension measurements would fall within $\mu \pm 3\sigma$; that is, from 146 to 404.

4.97 $c = 3.090$ (Looking at Table A, there appear to be three possible answers—3.08, 3.09, or 3.10. In fact, the answer is 3.090232....)

4.98 (a) Mean: 0.1; standard deviation: $\sqrt{\frac{p(1-p)}{400}} = 0.015$. **(b)** Approximately $N(0.1, 0.015)$. **(c)** Center: 0.1; control limits: 0.055 and 0.145. **(d)** This process is out of control. Points below the lower control limit would not be a problem here, but beginning with lot number 2, we see many points above the upper control limit, and every value of $\hat{p}$ is above the center line (with the exception of two points that fall *on* the center line). A failure rate above 0.1 is strongly indicated.

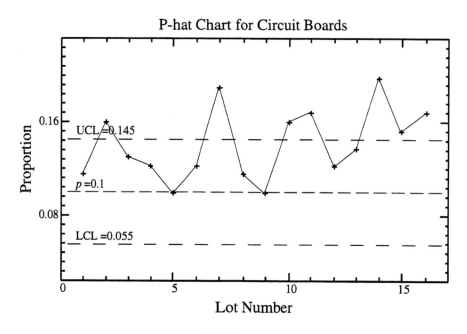

4.99 Center: p; control limits: $p \pm 3\sqrt{\frac{p(1-p)}{n}}$.

4.100 Center: 0.0225; control limits: −0.02724 and 0.07224. Since −0.02724 is a meaningless value for a proportion, the LCL may as well be set to 0, especially since we are concerned with the failure proportion being too high rather than too low.

Chapter Review

4.101 $P(\hat{p} > 0.50) = P(Z > \frac{0.50 - 0.45}{0.02225}) = P(Z > 2.247) = 0.01231$.

4.102 (a) $np = (25000)(0.141) = 3525$. **(b)** $P(X \geq 3500) = P(Z \geq \frac{3500 - 3525}{55.027}) = P(Z \geq -0.4543) = 0.6752$.

4.103 (a) $P(Z > \frac{105 - 100}{15}) = P(Z > \frac{1}{3}) = 0.36944$. **(b)** Mean: 100; standard deviation: 1.93649. **(c)** $P(Z > \frac{105 - 100}{1.93649}) = P(Z > 2.5820) = 0.00491$. **(d)** The answer to (a) could be quite different; (b) would be the same (it does not depend on normality at all). The answer we gave for (c) would be still be fairly reliable because of the central limit theorem.

4.104 $P(\frac{750}{12} < \overline{x} < \frac{825}{12}) = P(-1.732 < Z < 2.598) = 0.95368$.

4.105 (a) No—a count assumes only whole-number values, so it cannot be normally distributed. **(b)** $N(1.5, 0.02835)$. **(c)** $P(\overline{x} > \frac{1075}{700}) = P(Z > 1.2599) = 0.10386$.

4.106 (a) $\mu = np = 50$, $\sigma = \sqrt{np(1-p)} = \sqrt{40} = 6.3246$. **(b)** $P(X \geq 60) \doteq P(Z \geq 1.5811) = 0.0569$ (or find $P(\hat{p} \geq 0.24)$), using the fact that the mean of $\hat{p}$ is 0.2 and the standard deviation is 0.0253—the computation comes out the same after standardizing). For reference, the "exact" probability is 0.06885.

4.107 $P(5067 \text{ or more heads}) = P(\hat{p} \geq 0.5067) = P(Z \geq 1.34) = 0.0901$. If Kerrich's coin was "fair," we would see 5067 or more heads in about 9% of all repetitions of the experiment of flipping the coin 10,000 times, or about once every 11 attempts. This is *some* evidence against the coin being fair, but it is not by any means overwhelming.

CHAPTER 5 SOLUTIONS

Section 5.1

5.1 (a) 44% to 50%. **(b)** We do not have information about the whole population; we only know about a small sample. We expect our sample to give us a good estimate of the population value, but it will not be exactly correct. **(c)** The procedure used gives an estimate within 3 percentage points of the true value in 95% of all samples.

5.2 This is a statement about the *mean* score for all young men, not about individual scores. We are only attempting to estimate the center of the population distribution; the scores for individuals are much more variable. Also, "95%" is not a probability or a proportion; it is a confidence level.

5.3 (a) Mean: 280; standard deviation: 1.89737. **(b)** Below. **(c)** 2 standard deviations—3.8 points. **(d)** Below; the confidence intervals drawn may vary, of course. **(e)** 95%.

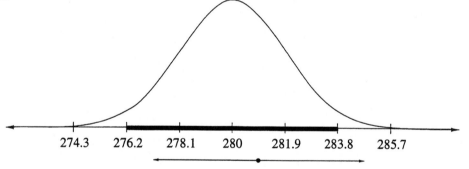

5.4 (a) $N(\mu, 0.05657)$. **(b)** See the sketch above. For this problem, the "numbers" below the axis would be $\mu - 0.16971$, $\mu - 0.11314$, $\mu - 0.05657$, μ, etc. **(c)** $m = 0.11314$ (2 standard deviations). **(d)** 95%. **(e)** See above.

5.5 11.78 ± 0.77, or 11.01 to 12.55 years.

5.6 (a) The stemplot is *somewhat* normal in appearance—not overwhelmingly so, but reasonably close. **(b)** 35.091 ± 4.272, or 30.819 to 39.363. **(c)** We base our confidence interval on the assumption that we have an SRS from the population. If all the students are in the same class, our methods are not reliable—that class might not be representative of the population of all third-graders in the district.

1	44
1	5899
2	2
2	55667789
3	13344
3	555589
4	0011234
4	5667789
5	1224

5.7 (a) The distribution is slightly skewed to the right.
(b) 224.002 ± 0.029, or 223.973 to 224.031.

2239	01
2239	66788889
2240	01
2240	589
2241	2

5.8 (a) 3.2 ± 0.329, or 2.871 to 3.529. **(b)** 3.2 ± 0.190, or 3.010 to 3.390.

5.9 (a) 0.8354 to 0.8454. **(b)** 0.8275 to 0.8533. **(c)** At right—increasing confidence makes the interval longer.

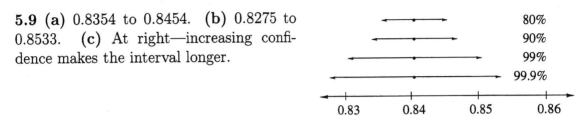

5.10 0.00506—which is half the margin of error with $n = 3$.

5.11 (a) 271.4 to 278.6. **(b)** 90%: 272.0 to 278.0; 99%: 270.3 to 279.7. **(c)** 90%: 3.0; 95%: 3.6; 99%: 4.7. Margin of error goes up with increasing confidence.

5.12 (a) 271.4 to 278.6. **(b)** 267.6 to 282.4. **(c)** 273.1 to 276.9. **(d)** 250: 7.4; 1077: 3.6; 4000: 1.9. Margin of error decreases with larger samples (by a factor of $\sqrt{n}$).

5.13 (a) 10.00209 to 10.00251. **(b)** 22 (21.64).

5.14 68 (67.95).

5.15 35 (34.57).

5.16 (a) The computations are correct. **(b)** Since the numbers are based on a voluntary response, rather than an SRS, the methods of this section cannot be used—the interval does not apply to the whole population.

5.17 (a) The interval was based on a method that gives correct results 95% of the time. **(b)** Since the margin of error was 2%, the true value of p could be as low as 49%. The confidence interval thus contains some values of p which give the election to Ford. **(c)** The proportion of voters that favor Carter is not random—either a majority favors Carter, or they don't. Discussing probabilities about this proportion has little meaning; the "probability" the politician asked about is either 1 or 0 (respectively).

5.18 (a) We can be 99% confident that between 63% and 69% of all adults favor such an amendment. We estimate the standard deviation of the distribution of $\hat{p}$ to be about $\sqrt{(0.66)(0.34)/1664} = 0.01161$; dividing 0.03 (the margin of error) by this gives $z^* = 2.58$, the critical value for a 99% confidence interval. **(b)** The survey excludes people without telephones (a large percentage of whom would be poor), so this group would be underrepresented. Also, Alaska and Hawaii are not included in the sample.

5.19 1.888 to 2.372.

5.20 (a) The intended population is hotel managers (perhaps specifically managers of hotels of the particular size range mentioned). However, because the sample came entirely from Chicago and Detroit, it may not do a good job of representing that larger population. There is also the problem of voluntary response. **(b)** 5.101 to 5.691. **(c)** 4.010 to 4.786. **(d)** We have a large enough sample size that the central limit theorem applies (if we accept the sample as an SRS).

5.21 (a) The intended population is "the American public"; the population which was actually sampled was "citizens of Indianapolis (with listed phone numbers)." **(b)** Food stores: 15.22 to 22.12; Mass merchandisers: 27.77 to 36.99; Pharmacies: 43.68 to 53.52. **(c)** The confidence intervals do not overlap at all; in particular, the *lower* confidence limit of the rating for pharmacies is higher than the *upper* confidence limit for the other stores. This indicates that the pharmacies are *really* higher.

5.22 $23,014 to $23,892.

5.23 505 residents.

5.24 The sample size for women was more that twice as large as that for men. Larger sample sizes lead to smaller margins of error (with the same confidence level).

5.25 $657.14 to $670.86. (The sample size is not used; the standard deviation given is σ_{estimate}).

Section 5.2

5.26 (a) $N(115, 6)$. **(b)** The actual result lies out toward the high tail of the curve, while 118.6 is fairly close to the middle. Assuming H_0 is true, observing a value like 118.6 would not be surprising, but 125.7 is less likely, and therefore provides evidence against H_0. **(c)** Below.

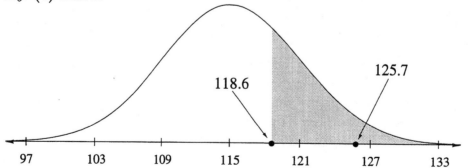

5.27 (a) $N(31\%, 1.518\%)$. **(b)** The lower percentage lies out in the low tail of the curve, while 30.2% is fairly close to the middle. Assuming H_0 is true, observing a value like 30.2% would not be surprising, but 27.6% is unlikely, and therefore provides evidence against H_0. **(c)** Below.

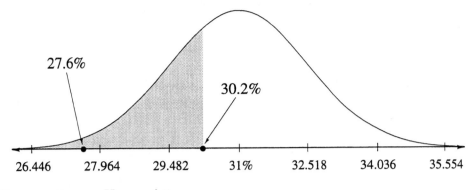

5.28 $H_0 : \mu = 5$ mm; $H_a : \mu \neq 5$ mm.

5.29 $H_0 : \mu = \$42,500$; $H_a : \mu > \$42,500$.

5.30 $H_0 : \mu = 50$; $H_a : \mu < 50$.

5.31 $H_0 : \mu = 2.6$; $H_a : \mu \neq 2.6$.

5.32 The P-values are 0.2743 and 0.0373, respectively.

5.33 The P-values are 0.2991 and 0.0125, respectively.

5.34 (a) $\bar{x} = 398$. **(b)** A $N(354, 19.053)$ density (below). **(c)** 0.0105. **(d)** It is significant at $\alpha = 0.05$, but not at $\alpha = 0.01$. This is pretty convincing evidence against H_0.

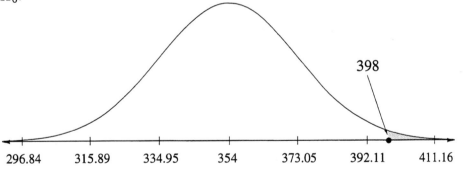

5.35 (a) A $N(0, 5.3932)$ density (below). **(b)** 0.1004. **(c)** Not significant at $\alpha = 0.05$. The study gives *some* evidence of increased compensation, but it is not very strong— it would happen 10% of the time just by chance.

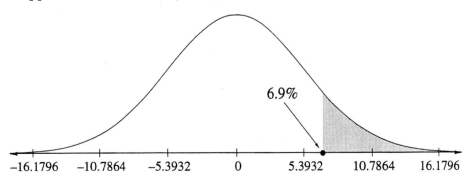

5.36 If church attenders are no more ethnocentric than nonattenders, then the outcomes observed for *this* sample would occur in less than 1 out of 20 instances. This being unlikely, we conclude that churchgoers are more ethnocentric.

5.37 Comparing men's and women's earnings for our sample, we observe a difference so large that it would only occur in 3.8% of all samples if men and women actually earned the same amount. Based on this, we conclude that men earn more.

While there is almost certainly *some* difference between earnings of black and white students in our sample, it is relatively small—if blacks and whites actually earn the same amount, we would still observe a difference as big as what we saw almost half (47.6%) of the time.

5.38 (a) $H_0 : \mu = 224$ vs. $H_a : \mu \neq 224$. (b) $z = 0.1292$. (c) $P = 0.8972$—this is reasonable variation when the null hypothesis is true, so we do not reject H_0.

5.39 (a) $H_0 : \mu = 300$ vs. $H_a : \mu < 300$. (b) $z = -0.7893$. (c) $P = 0.2150$—this is reasonable variation when the null hypothesis is true, so we do not reject H_0.

5.40 (a) $z = -2.200$. (b) Yes, because $|z| > 1.960$. (c) No, because $|z| < 2.576$.

5.41 (a) Yes, because $z > 1.645$. (b) Yes, because $z > 2.326$.

5.42 (a) 99.86 to 108.40. (b) Because 105 falls in this 90% confidence interval, we cannot reject $H_0 : \mu = 105$ in favor of $H_a : \mu \neq 105$.

5.43 (a) $N(0, 0.11339)$ (below). (b) $\overline{x} = 0.27$ lies out in the tail of the curve, while 0.09 is fairly close to the middle. Assuming H_0 is true, observing a value like 0.09 would not be surprising, but 0.27 is unlikely, and therefore provides evidence against H_0. (c) $P = 0.4274$ (the shaded region below). (d) $P = 0.0173$.

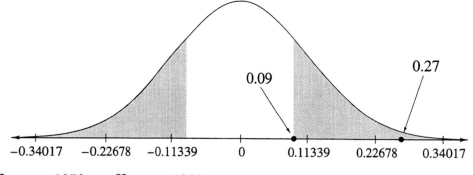

5.44 $H_0 : \mu = 1250$ vs. $H_a : \mu < 1250$.

5.45 $H_0 : \mu = 18$ vs. $H_a : \mu < 18$.

5.46 Hypotheses: $H_0 : \mu = -0.545$ vs. $H_a : \mu > -0.545$. Test statistic: $z = 1.957$. P-value: $P = 0.0252$. We conclude that the mean freezing point really is higher, and thus the supplier *is* apparently adding water.

5.47 (a) No, because $|z| < 1.960$. (b) No, because $|z| < 1.645$.

5.48 P is between 0.02 and 0.04 (in fact, $P = 0.0278$).

5.49 P is between 0.005 and 0.01 (in fact, $P = 0.0078$).

5.50 P is between 0.10 and 0.20 (in fact, $P = 0.1707$).

5.51 $P = 0.1292$. Although this sample showed *some* difference in market share between pioneers with patents or trade secrets and those without, the difference was small enough that it could have arisen merely by chance. The observed difference would occur in about 13% of all samples even if there is *no* difference between the two types of pioneer companies.

5.52 If there were no differences between brands, then what was observed in this particular sample occurs less than once in 1000 times. Since this is so unlikely, we conclude that perceived age does differ between brands. While only a sample of advertisements was used for this study, if it was a randomly chosen sample, it should be a fair representation of *all* ads for these brands.

5.53 When a test is significant at the 1% level, it means that if the null hypothesis is true, outcomes similar to those seen are expected to occur less than once in 100 repetitions of the experiment or sampling. "Significant at the 5% level" means we have observed something which occurs in less than 5 out of 100 repetitions (when H_0 is true). Something that occurs "less than once in 100 repetitions" also occurs "less than 5 times in 100 repetitions," so significance at the 1% level implies significance at the 5% level (or any higher level).

5.54 The explanation is not correct; either H_0 is true (in which case the "probability" that H_0 is true equals 1) or H_0 is false (in which case this "probability" is 0). "Statistically significant at the $\alpha = 0.05$ level" means that *if* H_0 is true, we have observed outcomes that occur less than 5% of the time.

Section 5.3

5.55 (a) $z = 1.64$; not significant at 5% level ($P = 0.0505$). **(b)** $z = 1.65$; significant at 5% level ($P = 0.0495$).

5.56 (a) $P = 0.3821$. **(b)** $P = 0.1714$. **(c)** $P = 0.0014$.

5.57 $n = 100$: 452.24 to 503.76. $n = 1000$: 469.85 to 486.15. $n = 10000$: 475.42 to 480.58.

5.58 No—the percentage was based on a voluntary response sample, and so cannot be assumed to be a fair representation of the population. Such a poll is likely to draw a higher-than-actual proportion of people with a strong opinion, esp. a strong negative opinion.

5.59 (a) No—in a sample of size 500, we expect to see about 5 people who have a "P-value" of 0.01 or less. These four *might* have ESP, or they may simply be among the "lucky" ones we expect to see. **(b)** The researcher should repeat the procedure on these four to see if they again perform well.

5.60 A test of significance answers question (b).

5.61 We might conclude that customers prefer design A, but perhaps not "strongly." Because the sample size is so large, this statistically significant difference may not be of any practical importance.

5.62 (a) 0.05. **(b)** Out of 77 tests, we can expect to see about 3 or 4 (3.85, to be precise) significant tests at the 5% level.

Section 5.4

5.63 (a) H_0 : the patient is ill (or "the patient should see a doctor"); H_a : the patient is healthy (or "the patient should not see a doctor"). A Type I error means a false negative—clearing a patient who should be referred to a doctor. A Type II error is a false positive—sending a healthy patient to the doctor. **(b)** One might wish to lower the probability of a false negative so that most ill patients are treated. On the other hand, if money is an issue, or there is concern about sending too many patients to see the doctor, lowering the probability of false positives might be desirable.

5.64 (a) Reject H_0 if $z < -2.326$. **(b)** 0.01 (the significance level). **(c)** We accept H_0 if $\bar{x} \geq 270.185$, so when $\mu = 270$, $P(\text{Type II error}) = P(\bar{x} \geq 270.185) = P(\frac{\bar{x}-270}{60/\sqrt{840}} \geq \frac{270.185-270}{60/\sqrt{840}}) = 0.4644$.

5.65 (a) 0.50. **(b)** 0.1841. **(c)** 0.0013.

5.66 (a) Reject H_0 if $\bar{x} \geq 0.5202$. **(b)** 0.9666.

5.67 (a) Reject H_0 if $\bar{x} \leq 297.985$, so the power against $\mu = 299$ is 0.2037. **(b)** The power against $\mu = 295$ is 0.9926. **(c)** The power against $\mu = 290$ would be greater—it is further from μ_0 (300), so it easier to distinguish from the null hypothesis.

5.68 (a) 0.5086. **(b)** 0.9543.

5.69 (a) Reject if $\bar{x} \geq 0.87011$ or $\bar{x} \leq 0.84989$. **(b)** Power: 0.89353. **(c)** $1 - 0.89353 = 0.10647$.

5.70 (a) We reject H_0 if $\bar{x} \geq 131.46$ or $\bar{x} \leq 124.54$. Power: 0.9246. **(b)** Power: 0.9246 (same as (a)). Over 90% of the time, this test will detect a difference of 6 (in either the positive or negative direction). **(c)** The power would be higher—it is easier to detect greater differences than smaller ones.

5.71 $P(\text{Type I error}) = 0.05$. $P(\text{Type II error}) = 1 - 0.9926 = 0.0074$.

5.72 Power: $1 - 0.4644 = 0.5356$.

5.73 A test having low power may do a good job of not incorrectly rejecting the null hypothesis, but it is likely to accept H_0 even when some alternative is correct, simply because it is difficult to distinguish between H_0 and "nearby" alternatives.

Chapter Review

5.74 (a) The plot is reasonably symmetric for such a small sample. **(b)** 26.06 to 34.74. **(c)** $H_0 : \mu = 25$ vs. $H_a : \mu > 25$; $z = 4.084$; P-value is essentially 0. (We knew from (b) that it had to be smaller than 0.025). This is strong evidence against H_0.

2	034
2	
3	01124
3	6
4	3

5.75 (a) 141.6 to 148.4. **(b)** $H_0 : \mu = 140$ vs. $H_a : \mu > 140$; $z = 2.421$; P-value is about 0.0077. This strongly supports H_a over H_0. **(c)** We must assume that the 15 cuttings in our sample are an SRS. Since our sample is not too large, the population should be normally distributed, or at least not extremely nonnormal.

5.76 12.285 to 13.515. This assumes that the babies are an SRS from the population. The population should not be too nonnormal (although a sample of size 26 will overcome quite a bit of skewness).

5.77 (a) $H_0 : \mu = 32$ vs. $H_a : \mu > 32$. **(b)** $z = 1.8639$; P-value is 0.0312. This is strong evidence against H_0—observations this extreme would only occur in about 3 out of 100 samples if H_0 were true.

5.78 (a) Narrower; lowering confidence level decreases the interval size. **(b)** Yes: \$32,000 falls outside the 99% confidence interval, indicating that $P < 0.01$.

5.79 (a) Margin of error decreases. **(b)** The P-value decreases (the evidence against H_0 becomes stronger). **(c)** The power increases (the test becomes better at distinguishing between the null and alternative hypotheses).

5.80 $H_0 : p = \frac{18}{38}$ vs. $H_a : p \neq \frac{18}{38}$.

5.81 No—"$P = 0.03$" *does* mean that the null hypothesis is unlikely, but only in the sense that the evidence (from the sample) would not occur very often if H_0 were true. P is a probability associated with the sample, not the null hypothesis; H_0 is either true or it isn't.

5.82 Yes—significance tests allow us to discriminate between random differences ("chance variation") that might occur when the null hypothesis is true, and differences that are unlikely to occur when H_0 is true.

5.83 (a) The difference observed in the study would occur in less than 1% of all samples if the two populations actually have the same proportion. **(b)** The interval is constructed using a method that is correct (i.e., contains the actual proportion) 95% of the time. **(c)** No—treatments were not randomly assigned, but instead were chosen by the mothers. Mothers who choose to attend a job training program may be more inclined to get themselves out of welfare.

CHAPTER 6 SOLUTIONS

Section 6.1

6.1 $37/\sqrt{4} = 18.5$.

6.2 (a) 2.015. **(b)** 2.518.

6.3 (a) 2.145. **(b)** 0.688.

6.4 (a) 2.262. **(b)** 2.861. **(c)** 1.440.

6.5 (a) 14. **(b)** 1.82 is between 1.761 ($p = 0.05$) and 2.145 ($p = 0.025$). **(c)** The P-value is between 0.025 and 0.05 (in fact, $P = 0.0451$). **(d)** $t = 1.82$ is significant at $\alpha = 0.05$ but not at $\alpha = 0.01$.

6.6 (a) 24. **(b)** 1.12 is between 1.059 ($p = 0.15$) and 1.318 ($p = 0.10$). **(c)** The P-value is between 0.30 and 0.20 (in fact, $P = 0.2738$). **(d)** $t = 1.12$ is not significant at either $\alpha = 0.10$ or at $\alpha = 0.05$.

6.7 (a) $\overline{x} = 1.75$ and $s = 0.1291$, so $SE(\overline{x}) = 0.06455$. **(b)** 1.598 to 1.902.

6.8 (a) $\overline{x} = 5.36667$ and $SE(\overline{x}) = 0.27162$. **(b)** 4.819 to 5.914.

6.9 $H_0 : \mu = 1.3$ vs. $H_a : \mu > 1.3$; $t = 6.9714$; the P-value is between 0.005 and 0.0025 (in fact, $P = 0.003$). This is very strong evidence against the null; we conclude that DDT does slow nerve recovery.

6.10 (a) μ is the difference between the population mean yields for Variety A plants and Variety B plants; that is, $\mu = \mu_A - \mu_B$. Another (equivalent) description is: μ is the mean difference between Variety A yields and Variety B yields. **(b)** $H_0 : \mu = 0$ vs. $H_a : \mu > 0$. **(c)** $t = 1.295$, $P = 0.1137$. This is not enough evidence to reject H_0—the difference could be due to chance variation.

6.11 (a) Randomly assign 12 (or 13) into a group which will use the right-hand knob first; the rest should use the left-hand knob first. Alternatively, for each student, randomly select which knob he or she should use first. **(b)** μ is the difference between right-handed times and left-handed times; the null hypothesis is $H_0 : \mu = 0$ (no difference). How the alternative is written depends on exactly how μ is defined. Let μ_R be the mean right-hand thread time for all right-handed people (or students), and μ_L be the mean left-hand thread time. As described above, we would most naturally write $\mu = \mu_R - \mu_L$; in this case, $H_a : \mu < 0$. Alternatively, we might define $\mu = \mu_L - \mu_R$, so that $H_a : \mu > 0$. Either way, the null hypothesis says $\mu_R = \mu_L$ and the alternative is

$\mu_R < \mu_L$. **(c)** $\overline{x} = -13.32$ (or $+13.32$), $SE(\overline{x}) = 4.5872$, $t = \pm 2.9037$, and $P = 0.0039$. We reject H_0 in favor of H_a.

6.12 5.47 to 21.17 seconds. For our sample $\overline{x}_R \div \overline{x}_L = 88.7\%$; this suggests that right-handed students working on an assembly line with right-handed threads would complete their task in about 90% of the time that it would take them to complete the same task with left-handed threads.

6.13 (a) 1.54 to 1.80. **(b)** We are told the distribution is symmetric; because the scores range from 1 to 5, there is a limit to how much skewness there might be. In this situation, the assumption that the 17 Mexicans are an SRS from the population is the most crucial.

6.14 (a) $H_0 : \mu = 0$ vs. $H_a : \mu > 0$. $t = 43.5$; the P-value is basically 0, so we reject H_0 and conclude that the new policy would increase credit card usage. **(b)** \$312 to \$352. **(c)** The sample size is very large, and we are told that we have an SRS. This means that outliers are the only potential snag, and there are none. **(d)** Make the offer to an SRS of 200 customers, and choose another SRS of 200 as a control group. Compare the mean increase for the two groups.

6.15 (a) The distribution is slightly skewed, but there are no apparent outliers. **(b)** $H_0 : \mu = 224$ vs. $H_a : \mu \neq 224$. $t = 0.12536$ and $P = 0.9019$, so we have very little evidence against H_0.

2239	01
2239	6688899
2240	002
2240	69
2241	02

6.16 (a) At right. **(b)** $H_0 : \mu = 105$ vs. $H_a : \mu \neq 105$, $t = -0.3195$, $P = 0.7554$. We do not reject the null hypothesis— the mean detector reading could be 105.

9	2
9	578
10	024
10	55
11	1
11	9
12	2

6.17 (a) Approximately 2.403 (from Table C), or 2.405 (using software). **(b)** Using $t^* = 2.403$: Reject H_0 if $t > 2.403$, which means $\overline{x} > 36.70$. **(c)** The power against $\mu = 100$ is 0.99998—basically 1. A sample of size 50 should be quite adequate.

6.18 (a) The power is 0.5287. (Reject H_0 if $t > 1.833$, i.e., if $\overline{x} > 0.4811$.) **(b)** The power is 0.9034. (Reject H_0 if $t > 1.711$, i.e., if $\overline{x} > 0.2840$.)

6.19 (a) 9. **(b)** $P = 0.0255$; it lies between 0.05 and 0.025.

6.20 (a) 21.47 to 26.53. **(b)** The sample is large enough that deviations from the assumptions do not greatly affect the validity of the t test.

6.21 (a) 111.22 to 118.58. **(b)** We assume that the 27 members of the placebo group can be viewed as an SRS of the population, and that the distribution of seated systolic BP in this population is normal, or at least not too nonnormal. Since the sample size is somewhat large, the proceduare should be valid as long as the data show no outliers and no strong skewness.

6.22 $\overline{x} = 22.125$. The standard error of $\overline{x}$ is 1.0451; the margin of error depends on the confidence level—two possible answers are ± 3.326 (if $C = 95\%$) and ± 6.105 (if $C = 99\%$). A margin of error of ± 3.326 means that we are 95% confident—that is, we have used a procedure that is correct 95% of the time—that the actual mean is between 18.799 and 25.451.

6.23 (a) Standard error of the mean. **(b)** $s = 0.01\sqrt{3} = 0.01732$. **(c)** 0.84 ± 0.0292; i.e., 0.8108 to 0.8692.

6.24 (a) $H_0 : \mu = 0$ vs. $H_a : \mu > 0$, where μ is the population mean of post-test minus pre-test scores. **(b)** Stemplot (right) shows no outliers, with a slight skewness to the left, but nothing too strong. **(c)** $\overline{x} = 1.45$, $SE(\overline{x}) = 0.71626$, and $t = 2.0244$; the P-value equals 0.057. This is not significant at either the 5% or 1% levels. **(d)** 0.2116 to 2.6884.

```
-0 | 54
-0 | 32
-0 | 11
 0 | 11
 0 | 2223333
 0 | 4455
 0 | 7
```

6.25 (a) $H_0 : \mu = 0$ vs. $H_a : \mu \neq 0$. For each subject, randomly choose which test to administer first. Alternatively, randomly assign 11 subjects to the "ARSMA first" group, and the rest to the "BI first" group. **(b)** $t = 4.27$; the P-value is less than 0.001, so we reject H_0. **(c)** 0.1292 to 0.3746.

6.26 (a) The sample size is large enough that skewness (in the absence of outliers) has little effect on our procedures. **(b)** df = 103. **(c)** 2.419 to 11.381; the data must be an SRS from the population of all corporate CEOs.

6.27 We know the data for *all* presidents; we know about the whole population, not just a sample. (We might want to try to make statements about future presidents, but doing so from this data would be highly questionable; they can hardly be considered an SRS from the population).

6.28 (a) 2.080. **(b)** Reject H_0 if $|t| \geq 2.080$, i.e., if $|\overline{x}| \geq 0.133$. **(c)** $P(|\overline{x}| \geq 0.133) = P(\overline{x} \leq -0.133 \text{ or } \overline{x} \geq 0.133) = P(Z \leq -5.207 \text{ or } Z \geq -1.047) = 0.852$.

Section 6.2

6.29 (a) (3)—two samples. **(b)** (2)—matched pairs.

6.30 (a) (1)—single sample. **(b)** (3)—two samples.

6.31 (a) $H_0 : \mu_1 = \mu_2$ vs. $H_a : \mu_1 < \mu_2$, where μ_1 is the beta-blocker population mean pulse rate and μ_2 is the placebo mean pulse rate. $t = -2.4525$; use a $t(29)$ distribution, which gives $P = 0.01022$. This makes the result significant at 5% but not at 1%. **(b)** -0.6311 to 10.8311.

6.32 $H_0 : \mu_1 = \mu_2$ vs. $H_a : \mu_1 > \mu_2$, where μ_1 and μ_2 are the mean number of beetles on untreated (control) plots and malathion-treated plots, respectively. $t = 5.8090$, which yields $P < 0.0001$ for a $t(12)$ distribution—this is significant at the 1% level.

6.33 (a) If k (the degrees of freedom) is reasonably large, the $t(k)$ distribution looks enough like the $N(0,1)$ distribution that for $t = 7.36$, we can conclude that the P-value is tiny (based on the 68–95–99.7 rule), so the result is significant. **(b)** Use a $t(32)$ distribution.

6.34 (a) Because the sample sizes are so large (and the sample sizes are almost the same), deviation from the assumptions have little effect. **(b)** Using $t^* = 1.660$ from a $t(100)$ distribution, the interval is \$412.68 to \$635.58. Using $t^* = 1.6473$ from a $t(620)$ distribution (obtained with software), the interval is \$413.54 to \$634.72. **(c)** The sample is not *really* random, but there is no reason to expect that the method used should introduce any bias into the sample. **(d)** Students without employment were excluded, so the survey results can only (possibly) extend to *employed* undergraduates. Knowing the number of unreturned questionnaires would also be useful.

6.35 (a) Both stemplots show no outliers; the experimental data (on the right side) is perhaps slightly skewed, but not enough to keep us from using the t procedures. **(b)** $H_0 : \mu_1 = \mu_2$ vs. $H_a : \mu_1 < \mu_2$ ($\mu_1 =$ control weight gain, ...). $\bar{x}_1 = 366.30$, $s_1 = 50.8052$, $\bar{x}_2 = 402.95$, $s_2 = 42.7286$, and $t = -2.469$; this gives $P = 0.0116$ (for a $t(19)$ distribution), so we reject H_0. **(c)** 5.58 to 67.72 grams.

87	2	
322	3	234
887666555	3	6899
3100	4	0011123333
66	4	578

6.36 (a) Women are on the left side; men on the right. Both distributions are slightly skewed to the right, and have one or two moderate high outliers. A t procedure may be (cautiously) used nonetheless, since the sum of the sample sizes is almost 40. **(b)** $H_0 : \mu_w = \mu_m$ vs. $H_a : \mu_w > \mu_m$. $\bar{x}_w = 141.056$, $s_w = 26.4363$, $\bar{x}_m = 121.250$, $s_m = 32.8519$, and $t = 2.0561$, so $P = 0.0277$ (for a $t(17)$ distribution). **(c)** For $\mu_m - \mu_w$: -36.57 to -3.05.

	7	05
	8	8
	9	12
931	10	489
5	11	3455
966	12	6
77	13	2
80	14	06
442	15	1
55	16	9
8	17	
	18	07
	19	
0	20	

6.38 (a) $H_0 : \mu_{\text{skilled}} = \mu_{\text{novice}}$ vs. $H_a : \mu_s > \mu_n$. **(b)** The t statistic we want is the "Unequal" value: $t = 3.1583$; its P-value is 0.0052. This is strong evidence against H_0. **(c)** Using $t^* = 1.833$ from a $t(9)$ distribution: 0.4922 to 1.8535. Using $t^* = 1.8162$ from a $t(9.8)$ distribution (from software): 0.4984 to 1.8473.

6.39 $H_0 : \mu_{\text{skilled}} = \mu_{\text{novice}}$ vs. $H_a : \mu_s \neq \mu_n$ (use a two-sided alternative since we have no preconceived idea of the direction of the difference). The t statistic we want is $t = 0.5143$; its P-value is 0.6165. There is no significant difference in weight between skilled and novice rowers.

6.40 Conservative: 19882 d.f.; more exact method: 38786 d.f. Either way, the distribution we use is almost exactly a $N(0,1)$ distribution. To three decimal places, $t^* = 2.576$, so the confidence interval is 27.91 to 32.09. Normality of SAT scores is not necessary since the sample sizes are so large.

6.41 (a) $H_0 : \mu_{\text{breast-fed}} = \mu_{\text{formula}}$ vs. $H_a : \mu_{bf} \neq \mu_f$ (the question implies no preconceived idea of the direction of the difference). Conservative: 18 d.f.; more exact method: 37.6 d.f. $t = 1.654$; the conservative P-value is 0.1155, and the more exact P-value is 0.1065; in either case, the results are not significant at the 10% level. **(b)** $t(18)$: $t^* = 2.101$; −0.2434 to 2.0434. $t(37.6)$: $t^* = 2.0251$; −0.2021 to 2.0021. **(c)** We assume that both groups are SRSs of mothers of both types, and that both distributions are normal—or at least don't depart too much from normality, and have no outliers. **(d)** This is not an experiment—the mothers chose the feeding method. This may confound the conclusions, since there may have been other factors that affected this choice.

6.42 (a) $H_0 : \mu_1 = \mu_2$ vs. $H_a : \mu_1 \neq \mu_2$. $t = -8.2379$; using either a $t(13)$ or a $t(21.8)$ distribution, the P-value is smaller than 0.0001, so there is a significant difference. **(b)** The fact that all the subjects in this study are college professors may have some confounding effects on the results. Additionally, all the subjects volunteered for a fitness program, which could bring in some further confounding.

6.43 (a) Use either a $t(52)$ or a $t(121.9)$ distribution. For the former, the nearest table value is $t^* = 2.009$, and the interval is -1.279 to 7.279. For 121.9 d.f., the interval is −1.216 to 7.216. **(b)** The samples taken by the market research firm do not give complete information about *all* stores. When we allow for the possible variation that might reasonably occur at the stores not included in the samples, we find that the actual sales might have dropped by 1.3 units, or could have risen by as much as 7.3 units.

6.44 (a) $H_0 : \mu_A = \mu_B$ vs. $H_a : \mu_A \neq \mu_B$; $t = -1.484$. Using $t(149)$ and $t(297.2)$ distributions, P equals 0.1399 and 0.1388, respectively; not significant in either case. The bank might choose to implement Proposal A even though the difference is not significant, since it may have a *slight* advantage over Proposal B. Otherwise, the bank should choose whichever option costs them less. **(b)** Because the sample sizes are equal and large, the t procedure is reliable in spite of the skewness. **(c)** This is an experiment—treatments are imposed by the bank. However, one other thing might be useful: statistics for a control group, to see if either plan increased spending.

6.45 (a) Stemplots show little skewness, but one moderate outlier (85) for the control group on the right. Nonetheless, the t procedures should be fairly reliable since the total sample size is 44. **(b)** $H_0 : \mu_t = \mu_c$ vs. $H_a : \mu_t < \mu_c$; $t = 2.311$. Using $t(20)$ and $t(37.9)$ distributions, P equals 0.0158 and 0.0132, respectively; reject H_0. **(c)** Randomization was not really possible, because existing classes were used—the researcher could not shuffle the students around.

	1	079
4	2	068
3	3	377
9964333	4	1222368
98776432	5	3455
721	6	02
1	7	
	8	5

6.46 (a) $H_0 : \mu_c = \mu_a$ vs. $H_a : \mu_c \neq \mu_a$; $t = 1.249$. Using $t(9)$ and $t(25.4)$ distributions, P equals 0.2431 and 0.2229, respectively; the difference is not significant. **(b)** –15.8 to 54.8 (using $t(9)$) or –12.6 to 51.6 (using $t(25.4)$). These intervals had to contain 0 because according to (a), the observed difference would occur in more than 22% of samples when the means are the same; thus 0 would appear in any confidence interval with a confidence level greater than 78%.

6.47 If they did (for example) 20 tests at the 5% level of significance, they might see 1 or 2 significant differences even when all null hypotheses are true.

6.48 (a) $t^* \doteq 2.364$, the value for a $t(100)$ distribution (since values for a $t(99)$ distribution are not given). **(b)** Reject H_0 when $\overline{x}_1 - \overline{x}_2 \geq 2.6746$. **(c)** Power: $P(Z \geq -2.0554) = 0.9801$.

6.49 (a) $H_0 : \mu_1 = \mu_2$ vs. $H_a : \mu_1 \neq \mu_2$; $t = (\overline{x}_1 - \overline{x}_2)/\sqrt{\frac{s_1^2}{n_1} + \frac{s_2^2}{n_2}}$. **(b)** $t^* \doteq 1.984$, from a $t(100)$ distribution. **(c)** Reject H_0 when $|\overline{x}_1 - \overline{x}_2| \geq 8.260$; power against $\mu_1 - \mu_2 = 10$: $P[(\overline{x}_1 - \overline{x}_2) \leq -8.260$ or $(\overline{x}_1 - \overline{x}_2) \geq 8.260] = P(Z \leq -4.386$ or $Z \geq -0.4179) = 0.662$.

6.50 (a) $H_0 : \mu_A = \mu_B$ vs. $H_a : \mu_A \neq \mu_B$; $t = (\overline{x}_A - \overline{x}_B)/\sqrt{\frac{s_A^2}{n_A} + \frac{s_B^2}{n_B}}$. **(b)** For a $t(349)$ distribution, $t^* = 1.967$; using a $t(100)$ distribution, take $t^* = 1.984$. **(c)** We reject H_0 when $|\overline{x}_A - \overline{x}_B| \geq 59.48$ (using $t^* = 1.967$). To find the power against $|\mu_A - \mu_B| = 100$, we choose *either* $\mu_A - \mu_B = 100$ or $\mu_A - \mu_B = -100$ (the probability is the same either way). Taking the former, we compute: $P[(\overline{x}_A - \overline{x}_B) \leq -59.48$ or $(\overline{x}_A - \overline{x}_B) \geq 59.48] = P(Z \leq -5.274$ or $Z \geq -1.340) = 0.9099$.

Repeating these computations with $t^* = 1.984$ gives power 0.9071.

Section 6.3

6.51 (a) $F^* = 3.68$. **(b)** Not significant at either 10% or 5% (in fact, $P = 0.1166$).

6.52 (a) Significant at 5%, but not at 1%. **(b)** Between 0.02 and 0.05 ($P = 0.0482$).

6.53 $F = 10.57$; P-value (for the two-sided alternative) is between 0.02 and 0.05 ($P = 0.0217$)—so this is significant at 5% but not at 1%.

6.54 (a) $H_0 : \sigma_{\text{skilled}} = \sigma_{\text{novice}}$ vs. $H_a : \sigma_s \neq \sigma_n$. **(b)** $F = 4.007$; P-value is between 0.05 and 0.10 ($P = 0.0574$).

6.55 (a) $H_0 : \sigma_s = \sigma_n$ vs. $H_a : \sigma_s \neq \sigma_n$. **(b)** $F = 2.196$; P-value is greater than 0.20 ($P = 0.2697$).

6.56 $F = 2.242$; the $F(19, 9)$ distribution doesn't show up in the tables, but by comparing to the $F(20, 9)$ distribution we see that the P-value is greater than 0.20 ($P = 0.2152$). The difference is not significant.

6.57 $F = 1.5443$; the P-value is 0.3725, so the difference is not statistically significant.

Chapter Review

6.58 (a) The observations are "before-and-after" weights, so the pairs of observations will be highly correlated—it is the change in weight that we are interested in. **(b)** We expect some variation in the weight change, and there may have been some loss due to chance, but the amount lost was so great that it is unlikely to occur merely by chance. In short, this weight-loss program seems to work. **(c)** Table C shows that the P-value must be smaller than 0.0005; in fact, it is less than 0.00002.

6.59 (a) This is a two-sample t test—the two groups of women are (presumably) independent. **(b)** Use a $t(44)$ distribution. **(c)** The sample sizes are large enough that nonnormality has little effect on the reliability of the procedure.

6.60 (a) Using a $t(1361)$ distribution, you get $1016.56 to $1069.44; using a $t(2669.1)$ distribution, you get almost the same interval: $1016.58 to $1069.42. **(b)** Skewness will have little effect because the sample sizes are very large.

6.61 (a) $H_0 : \mu_1 = \mu_2$ vs. $H_a : \mu_1 < \mu_2$; $t = -8.947$; P-value is basically 0 (however one chooses the degrees of freedom). We reject H_0 and conclude that the workers have higher output. **(b)** The t procedures are robust against skewness when the sample sizes are large. **(c)** Insertions for experienced workers have (approximately) a $N(\overline{x}_2, s_2)$ distribution; the 68–95–99.7 rule tells us that 95% of all workers can insert between $\overline{x}_2 - 2s_2 = 29.66$ and $\overline{x}_2 + 2s_2 = 44.99$ pins in the allotted time. (We might also choose $\overline{x}_2 \pm 1.96s_2$, but given the other approximations we are making, $\pm 2s_2$ is quite adequate).

6.62 (a) $H_0 : \mu_1 = \mu_2$ vs. $H_a : \mu_1 > \mu_2$. Use a matched-pairs t procedure, since the pairs of scores are related. **(b)** The stemplot is skewed and not symmetric, and the sample size is not large enough to overcome these violations of the assumptions.

```
-1 | 6
-1 |
-0 | 6555
-0 | 3
 0 | 1234
 0 | 55
```

6.63 Both stemplots are reasonably symmetrical, though the nitrite group (on the right) may be slightly left-skewed. There are no extreme outliers. $H_0 : \mu_c = \mu_n$ vs. $H_a : \mu_c > \mu_n$; $t = 0.8909$ and P equals 0.1902 (using a $t(29)$ distribution) or 0.1884 (with 56.8 d.f.). In either case, the difference is not significant.

```
              |  5 |  0
           6  |  5 |  6
              |  6 |  23
         765  |  6 |  5668
       42211  |  7 |  34
        9987  |  7 |  67779
      422111  |  8 |  011233
        8765  |  8 |  56689
       41000  |  9 |
              |  9 |  559
          30  | 10 |  2
```

6.64 (a) The stemplot shows some left-skewness; however, for such a small sample, the data are not unreasonably skewed. There are no outliers. **(b)** 54.78 to 64.40.

```
4 | 9
5 | 1
5 | 5
6 | 1334
6 | 55
```

6.65 (a) The stemplot is reasonably symmetrical given the small sample size. There are no outliers. **(b)** 903.23 to 912.27. **(c)** No, because 910 falls inside the 95% confidence interval.

```
89 | 3
89 | 57
90 | 1
90 | 566678
91 | 34
91 | 688
92 | 1
```

6.66 (a) "s. e." stands for standard error (of the mean). $\overline{x}_1 = 2821$, $s_1 = 435.58$; $\overline{x}_2 = 2844$, $s_2 = 437.30$; $\overline{x}_3 = 0.24$, $s_3 = 0.59397$; $\overline{x}_4 = 0.39$, $s_4 = 1.0021$. **(b)** No. $t = -0.3532$, and $P = 0.7248$ (using $t(82)$) or 0.7243 (using $t(173.9)$)—in either case, there is little evidence against H_0. **(c)** Not very significant—$t = -1.1972$, and $P = 0.2346$ (using $t(82)$) or 0.2334 (using $t(128.4)$). **(d)** 0.207 to 0.573. **(e)** -0.3119 to 0.0119 (using $t(82)$) or -0.3114 to 0.0114 (using $t(128.4)$).

6.67 No—you have information about all Indiana counties (not just a sample).

6.68 (a) $t = 1.897$, so $P = 0.0326$—not significant at 1%. **(b)** Use a matched-pairs design, with each soil specimen split in half and measured by each method. Then test $H_0 : \mu_1 = \mu_2$ vs. $H_a : \mu_1 \neq \mu_2$.

6.69 (a) $H_0 : \mu_1 = \mu_2$ vs. $H_a : \mu_1 > \mu_2$; $t = 1.1738$, so $P = 0.1265$ (using $t(22)$) or 0.123453 (using $t(43.3)$). Not enough evidence to reject H_0. **(b)** -14.57 to 52.57, or -13.64 to 51.64. **(c)** 165.53 to 220.47. **(d)** We are assuming that we have two SRSs from each population, and that underlying distributions are normal. It is unlikely that we have random samples from either population, especially among pets.

6.70 The stemplot looks fairly good; 48.8 is perhaps a moderate low outlier, but is not too far from the other observations. Our estimate is the mean, $\bar{x} = 5.4479$. The standard error of the mean is 0.0410; the margin of error depends on the confidence level chosen. Here are three possibilities:

Confidence Level	Confidence Interval	Margin of error
90%	(5.3781,5.5177)	0.0698
95%	(5.3639,5.5320)	0.0840
99%	(5.3345,5.5613)	0.1134

```
48 | 8
49 |
50 | 7
51 | 0
52 | 6799
53 | 04469
54 | 2467
55 | 03578
56 | 12358
57 | 59
58 | 5
```

6.71 Choice of confidence level may vary, of course; using 95% confidence we get: for abdomen skinfold: -15.5 to -11.5 (using $t(19)$), or -15.4 to -11.6 (using $t(103.6)$). For thigh skinfold: -12.95 to -9.65 (using $t(19)$), or -12.86 to -9.74 (using $t(106.4)$).

6.72 (a) $H_0 : \sigma_m = \sigma_w$ vs. $H_a : \sigma_m \neq \sigma_w$; $F = 1.16$, which gives $P = 0.3754$ (looking up values in Table D allows us to determine that $P > 0.2$). **(b)** No—F procedures are not robust against nonnormality, even with large samples.

6.73 We have 30 available observations; the distribution is right-skewed, with a high outlier of 123. A 95% confidence interval for the mean city particulate level is 54.067 ± 7.246, or 46.82 to 61.31. If we discard the outlier, we get 51.69 ± 5.57, or 46.12 to 57.26.

Confidence intervals should be used with caution here—the outlier makes t procedures suspect, and discarding the outlier may cause us to underestimate the actual mean.

```
 2 | 3
 3 | 4899
 4 | 12222455689
 5 | 017789
 6 | 089
 7 | 12
 8 | 26
 9 |
10 |
11 |
12 | 3
```

6.74 The distribution of differences (city minus rural) has 26 observations. There are two high outliers (15 and 18). If we naïvely use the t procedures in spite of the outliers, we get $\bar{x} = 2.192$, $s = 4.691$, and $t = 2.383$; the P-value for testing $H_0 : \mu_c = \mu_r$ vs. $H_a : \mu_c > \mu_r$ is 0.0125—pretty strong evidence against H_0. A 95% confidence interval for $\mu_c - \mu_r$ is 0.297 to 4.087.

If we throw out those outliers, the data seem to be more suitable for a t procedure. Using the other 24 observations, $\bar{x} = 1.00$, $s = 2.1059$, $t = 2.326$, and $P = 0.0146$—we can still reject H_0, even though we have removed from the data the two strongest individual pieces of evidence against H_0. A 95% confidence interval for $\mu_c - \mu_r$ is 0.111 to 1.889.

```
-0 | 32
-0 | 11110
 0 | 01111111
 0 | 2222222
 0 | 5
 0 | 7
 0 |
 1 |
 1 |
 1 | 5
 1 |
 1 | 8
```

6.75 The plot shows a strong positive linear relationship; there is only one observation—(51,69)—which deviates from the pattern slightly. The regression line $\hat{y} = -2.580 + 1.0935x$ explains $r^2 = 95.1\%$ of the variation in the data. When $x = 88$, we predict $\hat{y} = 93.65$.

The point (108,123) is a potentially influential observation (although it does not seem to deviate from the pattern of the other points). Computing the regression line without this point gives $\hat{y} = 1.963 + 0.9942x$ and $r^2 = 92.1\%$, and $\hat{y} = 89.45$ when $x = 88$.

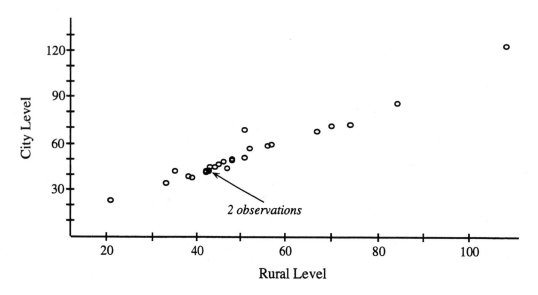

CHAPTER 7 SOLUTIONS

Section 7.1

7.1 (a) Population: the 175 residents of Tonya's dorm; p is the proportion who like the food. **(b)** $\hat{p} = 0.28$.

7.2 (a) The population is the 2400 students at Glen's college, and p is the proportion who believe tuition is too high. **(b)** $\hat{p} = 0.76$.

7.3 (a) The population is the 15,000 alumni, and p is the proportion who support the president's decision. **(b)** $\hat{p} = 0.38$.

7.4 (a) No—the population is not large enough relative to the sample. **(b)** Yes—we have an SRS, the population is 48 times as large as the sample, and the success count (38) and failure count (12) are both greater than 10. **(c)** No—there were only 5 or 6 "successes" in the sample.

7.5 (a) No—np_0 and $n(1 - p_0)$ are less than 10 (they both equal 5). **(b)** No—the expected number of failures is less than 10 ($n(1 - p_0) = 2$). **(c)** Yes—we have an SRS, the population is more than 10 times as large as the sample, and $np_0 = n(1 - p_0) = 10$.

7.6 (a) We have an SRS, the population is 50 times as large as the sample, and we observed 86 successes and 14 failures. **(b)** 0.792 to 0.928.

7.7 $\hat{p} = 0.3786$, $z = 2.72$, and $P = 0.0033$—reject $H_0 : p = \frac{1}{3}$ in favor of $H_a : p > \frac{1}{3}$.

7.8 (a) 59.4% to 72.6%. **(b)** $H_0 : p = 0.73$ vs. $H_a : p \neq 0.73$; $\hat{p} = 0.66$, $z = -2.23$ and $P = 0.02576$—reject H_0. **(c)** We have an SRS; assuming that there are at least 2000 new students, the population is more than 10 times as large as the sample; we observed 132 successes and 68 failures.

7.9 (a) $\hat{p} = 0.5005$, $z = 0.1549$, and $P = 0.8769$—accept the null hypothesis at any reasonable level of significance. **(b)** 0.4922 to 0.5088.

7.10 (a) 1051.7—round up to 1052. **(b)** 1067.1—round up to 1068; sixteen additional people.

7.11 450.2—round up to 451.

7.12 6.64% to 10.26%.

7.13 $\hat{p} = 0.05227$, $z = -3.337$, and $P < 0.0005$—very strong evidence against H_0, and in favor of $H_a : p < \frac{1}{10}$.

7.14 (a) 39.0% to 45.0%. **(b)** Since 50% falls outside the 99% confidence interval, this is strong evidence against $H_0 : p = 0.5$ in favor of $H_a : p < 0.5$. (In fact, $z = -6.75$ and P is tiny.) **(c)** 16589.4—round up to 16590. The use of $p^* = 0.5$ is reasonable because our confidence interval shows that the actual p is in the range 0.3 to 0.7.

7.15 (a) 0.0913 to 0.2060. **(b)** 304. **(c)** The sample comes from a limited area in Indiana, focuses on only one kind of business, and leaves out any businesses not in the Yellow Pages (there might be a few of these; perhaps they are more likely to fail). It is more realistic to believe that this describes businesses that match the above profile; it *might* generalize to food-and-drink establishments elsewhere, but probably not to hardware stores and other types of business.

7.16 (a) $H_0 : p = 0.5$ vs. $H_a : p > 0.5$, $z = 1.697$, $P = 0.0448$—reject H_0 at the 5% level. **(b)** 0.5071 to 0.7329. **(c)** The coffee should be presented in random order—some should get the instant coffee first, and others the fresh-brewed first.

7.17 (a) 4719. **(b)** 0.01125.

7.18 (a)

p	0.1	0.2	0.3	0.4	0.5	0.6	0.7	0.8	0.9
m	.0588	.0784	.0898	.0960	.0980	.0960	.0898	.0784	.0588
(b) m	.0263	.0351	.0402	.0429	.0438	.0429	.0402	.0351	.0263

The new margins of error are less than half their former size (in fact, they have decreased by a factor of $\frac{1}{\sqrt{5}} \doteq 0.447$).

Section 7.2

7.19 (a) −0.0208 to 0.1476. **(b)** The population-to-sample ratio is certainly large enough, and the smallest count in any category is 75—much larger than 5.

7.20 (a) 0.0341 to 0.0757. **(b)** All counts are larger than 5 (the smallest is 54), and the populations are much larger than the samples.

7.21 (a) $H_0 : p_1 = p_2$ vs. $H_a : p_1 \neq p_2$. The population-to-sample ratio is large enough, and the smallest number of people in any category is 94 (Catholics answering "Yes"). **(b)** $\hat{p}_1 = 0.6030$, $\hat{p}_2 = 0.5913$, $\hat{p} = 0.5976$; $z = 0.2650$, and $P = 0.7910$—H_0 is quite plausible given this sample.

7.22 (a) $H_0 : p_1 = p_2$ vs. $H_a : p_1 \neq p_2$; the populations are much larger than the samples, and 44 (the smallest count) is much bigger than 5. **(b)** $\hat{p} = 0.0933$, $z = -3.802$, and $P < 0.0002$—the difference is statistically significant.

7.23 (a) $H_0 : p_1 = p_2$ vs. $H_a : p_1 \neq p_2$; the populations are much larger than the samples, and 17 (the smallest count) is greater than 5. **(b)** $\hat{p} = 0.0632$, $z = 2.926$, and $P = 0.0034$—the difference is statistically significant. **(c)** Neither the subjects nor the researchers who had contact with them knew which subjects were getting which drug—if anyone had known, they might confound the outcome by letting their expectations or biases affect the results.

7.24 (a) $H_0 : p_1 = p_2$ vs. $H_a : p_1 \neq p_2$; $P = 0.0028$—the difference is statistically significant. **(b)** 0.1172 to 0.3919.

7.25 $H_0 : p_1 = p_2$ vs. $H_a : p_1 \neq p_2$; $P = 0.6981$—insufficient evidence to reject H_0.

7.26 $H_0 : p_1 = p_2$ vs. $H_a : p_1 \neq p_2$; $P = 0.9805$—insufficient evidence to reject H_0.

7.27 (a) $H_0 : p_1 = p_2$ vs. $H_a : p_1 > p_2$; $P = 0.0335$—reject H_0 (at the 5% level). **(b)** -0.0053 to 0.2336.

7.28 (a) 0.1626 to 0.2398. b(b) Zero does not lie in the 99% confidence interval for $p_2 - p_1$, so we would reject $H_0 : p_2 - p_1 = 0$ in favor of the two-sided alternative at the 1% level. **(c)** Yes—$P < 0.00002$.

7.29 The population-to-sample ratio is large enough, and the smallest count is 10—twice as big as it needs to be to allow the z procedures. Fatal heart attacks: $z = -2.67, P = 0.0076$. Non-fatal heart attacks: $z = -4.58, P < 0.000005$. Strokes: $z = 1.43, P = 0.1525$. The proportions for both kinds of heart attacks were significantly different; the stroke proportions were not.

7.30 (a) 0.2465 to 0.3359—since 0 is not in this interval, we would reject $H_0 : p_1 = p_2$ at the 1% level (in fact, P is practically 0). **(b)** No: $t = -0.8658$, which gives a P-value close to 0.4.

7.31 (a) -9.91% to -4.09%. Since 0 is not in this interval, we would reject $H_0 : p_1 = p_2$ at the 1% level. **(b)** -0.7944 to -0.4056. Since 0 is not in this interval, we would reject $H_0 : \mu_1 = \mu_2$ at the 1% level.

7.32 (a) $\hat{p}_1 = 0.1415$, $\hat{p}_2 = 0.1667$; $P = 0.6981$. **(b)** $P = 0.0336$. **(c)** From (a): -0.1056 to 0.1559. From (b): 0.001278 to 0.049036. The larger samples make the margin of error (and thus the length of the confidence interval) smaller.

Chapter Review

7.33 No—the data is not based on an SRS, and thus the z procedures are not reliable in this case. In particular, a voluntary response sample is typically biased.

7.34 No—$\hat{p} = 49.19\%$, $z = 0.1834$, and $P = 0.4272$.

7.35 92.3% to 93.7%.

7.36 (a) -0.1965 to -0.0210. **(b)** Yes—0 falls outside the confidence interval. In fact, $P = 0.0105$.

7.37 (a) No—$P = 0.1849$. **(b)** Yes—$P = 0.0255$. **(c)** The population is (presumably) very large, so the ratio of population-to-sample is big enough. Also, all the counts—45 and 29 in (a); 21, 7, 24, and 22 in (b)—are bigger than 5. The counts for baseball players are too small.

7.38 (a) The numbers are percentages of total revenues, not percentages of businesses in the sample. Specifically, for each business, a variable ("percentage of revenues from highly regulated businesses") was measured, and the average of these values was computed. **(b)** Use the t procedures from Chapter 6 (we would also need to know the standard deviations).

CHAPTER 8 SOLUTIONS

8.1 (a) 2×3. **(b)** 55.0%, 74.7%, and 37.5%. Some (but not too much) time spent in extracurricular activities seems to be beneficial. **(c)** At right. **(d)** Below. **(e)** The first and last columns have lower numbers than we expect in the "passing" row (and higher numbers in the "failing" row), while the middle column has this reversed—more passed than we would expect if all proportions were equal.

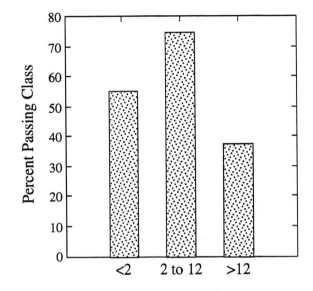

	<2	2–12	>12
C or better	13.78	62.71	5.51
D or F	6.22	28.29	2.49

8.2 (a) 3×2. **(b)** 22.5%, 18.6%, and 13.9%. A student's likelihood of smoking increases when one parent smokes, and increases even more when both smoke. **(c)** See exercise 2.67. **(d)** The null hypothesis says that parents' smoking habits have no effect on their children. **(e)** Below. **(f)** In column 1, row 1, the expected count is much smaller than the actual count; meanwhile, the actual count is lower than expected in the lower left. This agrees with what we observed before: children of non-smokers are less likely to smoke.

	Student smokes	Student does not smoke
Both parents smoke	332.49	1447.51
One parent smoke	418.22	1820.78
Neither parent smokes	253.29	1102.71

8.3 (b) $P = 0.0313$. Rejecting H_0 means that we conclude that there is a relationship between hours spent in extracurricular activities and performance in the course. **(c)** The highest contribution comes from column 3, row 2 (">12 hours of extracurricular activities, D or F in the course"). Too much time spent on these activities seems to hurt academic performance. **(d)** No—this study demonstrates association, not causation. Certain types of students may tend to spend a moderate amount of time in extracurricular activities and also work hard on their classes—one does not necessarily cause the other.

8.4 (b) P is essentially 0. By rejecting H_0, we conclude that there is a relationship between parents' smoking habits and those of their children. **(c)** The highest contributions come from C1 R1 ("both parents smoke, student smokes") and C1 R3 ("neither parent smokes, student smokes"). When both parents smoke, their student is much more likely to smoke; when neither parent smokes, their student is unlikely to smoke. **(d)** No—this study demonstrates association, not causation. There may be other factors (heredity or environment, for example) that cause *both* student and parent(s) to smoke.

8.5 (a) $(r-1)(c-1) = (2-1)(3-1) = 2$. **(b)** $X^2 = 6.926$ lies between 5.99 and 7.38; therefore, P is between 0.05 and 0.025.

8.6 (a) $(r-1)(c-1) = (3-1)(2-1) = 2$. **(b)** $X^2 = 37.568$ lies above 15.20 (the largest value in that row); therefore, $P < 0.0005$.

8.7 (a) and **(b)** Tables below. 25% (2 out of 8) of the expected counts are less than 5, which goes against our guidelines. **(c)** 3 d.f.; P-value is between 0.0025 and 0.001 (in fact, $P = 0.0018$). **(d)** Students with high goals show a higher proportion of passing grades than those who merely wanted to pass.

	Grade Received			
	Actual		Expected	
	$\geq$ C	D/F	$\geq$ C	D/F
Wanted $\geq$ C	5	9	9.65	4.35
Wanted $\geq$ B	41	23	44.10	19.90
Wanted an A	27	3	20.67	9.33
Wanted an "A+"	9	2	7.58	3.42

8.8 (a) 2.96%, 13.07%, and 6.36%. **(b)** and **(c)** Table below—actual counts above, expected counts below. Expected counts are all greater than 5, so the chi-square test is safe. **(d)** H_0 : there is no relationship between a member complaining and leaving the HMO vs. H_a : there is some relationship. 2 d.f.; $P < 0.0005$ (basically 0). **(e)** Members who file complaints—especially medical complaints—are more likely to leave the HMO.

	No Complaint	Medical Complaint	Non-medical Complaint
Stayed	721	173	412
	702.14	188.06	415.80
Left	22	26	28
	40.86	10.94	24.20

8.9 (a) 7.01%, 14.02%, and 13.05%. **(b)** and **(c)** Table below—actual counts above, expected counts below. Expected counts are all much bigger than 5, so the chi-square test is safe. H_0 : there is no relationship between worker class and race vs. H_a : there is some relationship. **(d)** 2 d.f.; $P < 0.0005$ (basically 0). **(e)** Black female child-care workers are more likely to work in non-household or preschool positions.

	Black	Other
Household	172	2283
	242.36	2212.64
Non-household	167	1024
	117.58	1073.42
Teachers	86	573
	65.06	593.94

8.10 (a) $H_0 : p_1 = p_2$ vs. $H_a : p_1 \neq p_2$. $z = -0.5675$ and $P = 0.5704$. (b) Table below. $X^2 = 0.322$, which equals z^2. With 1 d.f., Table E tells us that $P > 0.25$; a statistical calculator gives $P = 0.5704$. (c) Gastric freezing is not significantly more (or less) effective than a placebo treatment.

	Improved	Did Not Improve
Gastric Freezing	28	54
	29.73	52.28
Placebo	30	48
	28.27	49.72

Chapter Review

8.11 (a) $X^2 = 10.827$ (3 d.f.); $P = 0.0127$, which is significant at the 5% level, so we reject H_0. (b) Graph and table on right. The biggest difference between women and men is in Administration: a higher percentage of women chose this major. Meanwhile, a greater proportion of men chose other fields, especially Finance. (c) The largest chi-square components are the two from the "Administration" row. Many more women than we expect (91 actual, 76.36 expected) chose this major, while only 40 men chose this (54.64 expected). (d) The "Economics" row had expected counts of 6.41 and 4.59, respectively. Only the second number is less than 5, which is only one eighth (12.5%) of the counts in the table—the chi-square procedure is acceptable. (e) 386 responded, so 46.54% did not respond.

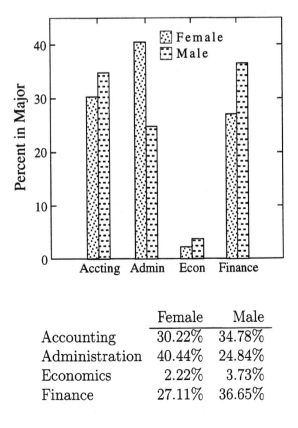

	Female	Male
Accounting	30.22%	34.78%
Administration	40.44%	24.84%
Economics	2.22%	3.73%
Finance	27.11%	36.65%

8.12 $X^2 = 3.277$ (4 d.f.); $P = 0.4874$, so we can accept H_0 : all types of companies had the same response rate.

	Response	Nonresponse
Metal Products	17	168
Machinery	35	266
Electrical Equipment	75	477
Transportation Equipment	15	85
Precision Instruments	12	78

8.13 (a) $H_0 : p_1 = p_2$, where p_1 and p_2 are the proportions of women customers in each city. $\hat{p}_1 = 0.8423$, $\hat{p}_2 = 0.6881$, $z = 3.9159$, and $P = 0.00009$. **(b)** $X^2 = 15.334$, which equals z^2. With 1 d.f., Table E tells us that $P < 0.0005$; a statistical calculator gives $P = 0.00009$. **(c)** 0.0774 to 0.2311.

8.14 4 degrees of freedom; $P > 0.25$ (in fact, $P = 0.4121$). There is not enough evidence to reject H_0 at any reasonable level of significance; the difference in the two income distributions is not statistically significant.

8.15 (a) $H_0 : p_1 = p_2$ vs. $H_a : p_1 < p_2$. The z test must be used because the chi-square procedure will not work for a one-sided alternative. **(b)** $z = -2.8545$ and $P = 0.0022$. Reject H_0; there is strong evidence in favor of H_a.

8.16 (a) $X^2 = 14.863$ with 2 d.f.; $P = 0.0006$, so reject H_0. Newark has a relatively high percentage (57.5%), Camden is very low (28.7%), and South Chicago falls in the middle (44.9%). **(b)** The data should come from independent SRSs of the (unregulated) child-care providers in each city.

8.17 H_0 : all proportions are equal vs. H_a : some proportions are different. Table on right. $X^2 = 10.619$ with 2 d.f., and $P = 0.0049$—good evidence against H_0, so we conclude that contact method makes a difference in response.

	Yes	No
Phone	168	632
One-on-one	200	600
Anonymous	224	576

8.18 (a) This is a 4×4 table, so there are 9 d.f. **(b)** $X^2 = 256.8$ (3 d.f.); P is essentially 0. The highest response rates occur from September to mid-April; the lowest occur in the summer months, when more people are likely to be on vacation (that is when the "No answer" percentage is highest).

8.19 H_0 : all refusal proportions are equal. The smallest expected count is 81.67, so the chi-square test is safe. $X^2 = 11.106$ (3 d.f.) and $P = 0.0112$—pretty strong evidence against H_0. The two largest components of X^2 are the July–Aug refusals (which were higher than expected) and the Jan–April refusals (which were low). Perhaps more people had "other things to do" in the summer, and had less pressing business during the winter.

	Refused	Other
Jan 1 to Apr 13	67	1491
	81.67	1476.33
Apr 21 to Jun 20	86	1503
	83.29	1505.71
Jul 1 to Aug 31	135	1940
	108.77	1966.23
Sept 1 to Dec 1	124	2514
	138.28	2499.72

8.20 (a) Yes, the evidence is *very* strong that a higher proportion of men die ($X^2 = 332.205$, 1 d.f.). Possibly many sacrificed themselves out of a sense of chivalry ("women and children first"). (b) For women, $X^2 = 103.767$ (2 d.f.)—a very significant difference. Over half of the lowest-status women died, but this percentage drops sharply when we look at middle-status women, and it drops again for high-status women. (c) For men, $X^2 = 34.621$ (2 d.f.)—another very significant difference (though not quite so strong as the women's value). Men with the highest status had the highest proportion surviving (over one-third). The proportion for low-status men was only about half as big, while middle-class men fared worst (only 12.8% survived).

8.21 (a) No; $X^2 = 1.051$ with 2 d.f., which gives $P = 0.5913$. (b) ABC News: $z = -0.7698$; $P = 0.4414$ (not significant). USA Today/CNN: $z = -5.1140$; P is essentially 0 (significant). New York Times/CBS: $z = -4.4585$; P is essentially 0 (significant). (c) An individual test will be wrong for only 5% of all samples. Imagine doing three tests in a row: Assuming the first test comes out correct (which it does 95% of the time), there is still a 5% chance that the next test will come out wrong, etc. Altogether, all three will be correct only 85.7%($= 0.95^3$) of the time.

8.22 (a) $X^2 = 2.186$ (1 d.f.); $P = 0.1393$—the difference is not statistically significant. (b) For "good condition" patients, $X^2 = 0.289$, while for "poor condition" patients, $X^2 = 0.019$ (both with 1 d.f.)—neither of these indicate significant differences ($P = 0.5909$ and 0.8904, respectively). (c) Though the effects are not statistically significant, Simpson's paradox is evident in that the P-value for the combined data is considerably lower than either of the P-values for the separate data.

8.23 (a) 0.5432 to 0.5968. (b) No: $z = 0.4884$, $P = 0.6253$. Or, using the exact counts (527/924 national, 96/174 student), $z = 0.4548$, $P = 0.6492$—again, not at all significant. (c) Yes: $z = 7.9215$, P is essentially 0. (d) Yes: $X^2 = 13.847$ with 2 d.f.; $P < 0.001$. Both student groups were less likely to believe that the military was censoring the news.

	Yes	No
National	702	222
	675.37	248.63
Student Grp 1	117	57
	127.18	46.82
Student Grp 2	129	70
	145.45	53.55

8.24 (a) The null hypothesis would say that there is no relationship between rows and columns; this would mean, for example, that knowing a certain substance causes cancer in mice tells us nothing about whether or not it causes cancer in rats. **(b)** Agreement between mice and rats means high numbers in the upper left and lower right, and low numbers in the other two entries. When we check for "agreement" of two proportions using a chi-square test, we might expect to see (for example) high numbers in the first column and low numbers in the second column, or vice versa. **(c)** Mice and rats agree on 84.7% (211/249) chemicals. 84.1% (111/132) of chemicals that test positive for mice also get a + from rats, while 85.5% (100/117) of chemicals that test negative for mice also get a − from rats.

CHAPTER 9 SOLUTIONS

9.1 **(a)** The distribution for perch has an extreme high outlier (20.9); the bream distribution has a mild low outlier (12.0). Otherwise there is no strong skewness. **(b)** Bream: 12.0 13.6 14.1 14.9 15.5; Perch: 13.2 15.0 15.55 16.675 20.9; Roach: 13.3 13.925 14.65 15.275 16.1. The most important difference seems to be that perch are larger than the other two fish. It also appears that (typically) bream *may* be slightly smaller than roach.

Bream		Perch		Roach	
12	0	12		12	
12		12		12	
13	333344	13	2	13	3
13	5677788889	13	69	13	6799
14	111233	14	3	14	0013
14	78899	14	55667889	14	677
15	001113	15	000000011112344	15	122344
15	5	15	56788999	15	6
16		16	0112333	16	1
16		16	8	16	
17		17	003	17	
17		17	56667789	17	
18		18	1	18	
18		18		18	
19		19		19	
19		19		19	
20		20		20	
20		20	9	20	

9.2 **(a)** $H_0 : \mu_1 = \mu_2 = \mu_3$—the mean weights of the three types are equal. **(b)** $F = 29.92$ and $P < 0.001$. **(c)** Perch seem to be actually different in weight from the other two species; the mean weight of bream and roach may not differ greatly (the two confidence intervals overlap).

9.3 **(a)** Mean yields: 131.03, 143.15, 146.23, 143.07, 134.8. The mean yields first increase with plant density, then decrease; the greatest yield occurs at or around 20,000 plants per acre. **(b)** $H_0 : \mu_1 = \mu_2 = \mu_3 = \mu_4 = \mu_5$ (all plant densities give the same mean yield per acre) vs. H_a : not all means are the same. **(c)** $F = 0.50$ and $P = 0.736$. The differences are not significant. **(d)** The sample sizes were small, which means there is a lot of potential variation in the outcome.

12,000		16,000		20,000		24,000		28,000	
11	38	11		11		11		11	9
12		12	1	12		12		12	
13		13	5	13	0	13	58	13	
14	3	14		14	0	14		14	
15	0	15	0	15	0	15	6	15	1
16		16	7	16	5	16		16	

9.4 (a) I, the number of populations, is 3; the sample sizes from the 3 populations are $n_1 = 35$, $n_2 = 55$ (after discarding the outlier), and $n_3 = 20$; the total sample size is $N = 110$. **(b)** numerator ("factor"): $I - 1 = 2$, denominator ("error"): $N - I = 107$. **(c)** Since $F > 7.41$, the largest critical value for an $F(2, 100)$ distribution in Table D, we conclude that $P < 0.001$.

9.5 (a) I, the number of populations, is 5; the sample sizes are $n_1 = 4$, $n_2 = 4$, $n_3 = 4$, $n_4 = 3$, and $n_5 = 2$; the total sample size is $N = 17$. **(b)** numerator ("factor"): $I - 1 = 4$, denominator ("error"): $N - I = 12$. **(c)** Since $F < 2.48$, the smallest critical value for an $F(4, 12)$ distribution in Table D, we conclude that $P > 0.100$.

9.6 (a) Populations: tomato varieties; response variable: yield. $I = 4$, $n_1 = n_2 = n_3 = n_4 = 10$, and $N = 40$; 3 and 36 d.f. **(b)** Populations: customers (responding to different package designs); response variable: attractiveness rating. $I = 6$, $n_1 = n_2 = n_3 = n_4 = n_5 = n_6 = 120$, and $N = 720$; 5 and 714 d.f. **(c)** Populations: dieters (under different diet programs); response variable: weight change after six months. $I = 3$, $n_1 = n_2 = 10$, $n_3 = 12$, and $N = 32$; 2 and 29 d.f.

9.7 Yes: $\dfrac{\text{largest } s}{\text{smallest } s} = \dfrac{1.186}{0.770} = 1.54$.

9.8 Yes: $\dfrac{\text{largest } s}{\text{smallest } s} = \dfrac{22.27}{11.44} = 1.95$.

9.9 (a) The biggest difference is that single men earn considerably less than men who have been or are married. Widowed and married men earn the most; divorced men earn about \$1300 less (on the average), and single men are \$4000 below that. **(b)** Yes: $\frac{8119}{5731} = 1.42$. **(c)** 3 and 8231. **(d)** The sample sizes are so large that even small differences would be found to be significant; we have some fairly large differences. **(e)** No—single men are likely to be younger than men in the other categories. This means that typically they have less experience, and have been with their companies less time than the others, and so have not received as many raises, etc.

9.10 (a) $H_0 : \mu_{r1} = \mu_{r2} = \mu_{r3}$ (all class rank means are same) vs. H_a : not all means are the same. **(b)** 2 and 253 (for all three tests). **(c)** Yes: $\frac{10.8}{10.5} = 1.03$, $\frac{1.31}{1.17} = 1.12$, and $\frac{.55}{.40} = 1.375$. Comparing to $F(2, 200)$ critical values, we find $P_{\text{rank}} > 0.100$, P_{sem} is between 0.025 and 0.010, and $P_{\text{grade}} < 0.001$. **(d)** Mean high school class rank varies little between the groups. Regarding the other two variables, there appears to be little difference between the CS and Sci/Eng majors. However, "semesters of HS math" and "average grade in HS math" both show a significant difference between CS/Sci/Eng majors and those in the "Other" category: on the average, the first two groups had about one half-semester more math, and had grades about 0.25 higher.

9.11 (a) MSE $= \frac{1}{107}\left[(34)(0.770)^2 + (54)(1.186)^2 + (19)(0.780)^2\right] = \frac{107.67}{107} = $ 1.0063; $s_p = \sqrt{\text{MSE}} = 1.003$. **(b)** Use $t^* = 1.984$ from a $t(100)$ distribution (since $t(107)$ is not available): $15.747 \pm t^* s_p/\sqrt{55} = 15.479$ to 16.015. Using software, we find that for a $t(107)$ distribution, $t^* = 1.982$; this rounds to the same interval.

9.12 $\overline{x} = \frac{1}{110}\left[(35)(14.131) + (55)(15.747) + (20)(14.605)\right] = 15.025$. MSG $= \frac{1}{2}\left[(35)(14.131 - 15.025)^2 + (55)(15.747 - 15.025)^2 + (20)(14.605 - 15.025)^2\right] = 30.086$. $F = \dfrac{30.086}{1.003} = 29.996$—reasonably close to Minitab's output.

9.13 MSE $= \frac{1}{12}\left[(3)(18.09)^2 + (3)(19.79)^2 + (3)(15.07)^2 + (2)(11.44)^2 + (1)(22.27)^2\right] = \dfrac{3595.7}{12} = 299.64$—this agrees with Minitab's output (except for roundoff error). $\overline{x} = \frac{1}{17}\left[(4)(131.03) + (4)(143.15) + (4)(146.23) + (3)(143.07) + (2)(134.75)\right] = 140.02$. MSG $= \frac{1}{4}\left[(4)(131.03 - 140.02)^2 + (4)(143.15 - 140.02)^2 + (4)(146.23 - 140.02)^2 + (3)(143.07 - 140.02)^2 + (2)(134.75 - 140.02)^2\right] = \frac{600.18}{4} = 150.04$.

9.14 Use $t^* = 1.782$ from a $t(12)$ distribution: 130.81 to 161.65.

Chapter Review

9.15 (a) Populations: nonsmokers, moderate smokers, and heavy smokers; response variable: hours of sleep per night. $I = 3$, $n_1 = n_2 = n_3 = 200$, and $N = 600$; 2 and 597 d.f. **(b)** Populations: different concrete mixtures; response variable: strength. $I = 5$, $n_1 = \cdots = n_5 = 6$, and $N = 30$; 4 and 25 d.f. **(c)** Populations: teaching methods; response variable: test scores. $I = 4$, $n_1 = n_2 = n_3 = 10$, $n_4 = 12$, and $N = 42$; 3 and 38 d.f.

9.16 (a) The data suggest that the presence of too many nematodes reduces growth. Table on right; stemplots below. **(b)** $H_0 : \mu_1 = \cdots = \mu_4$ (all mean heights are the same) vs. H_a : not all means are the same. This ANOVA tests whether nematodes affect mean

Nematodes	n_i	$\overline{x}_i$	s_i
0	4	10.650	2.053
1000	4	10.425	1.486
5000	4	5.600	1.244
10000	4	5.450	1.771

plant growth. **(c)** Minitab output is shown below. The first two levels (0 and 1000 nematodes) do not appear to be significantly different, nor do the last two. However, it does appear that somewhere between 1000 and 5000 nematodes, the tomato plants begin to feel the effects of the worms, and are hurt by their presence.

	0		1000		5000		10,000
3		3		3		3	2
4		4		4	6	4	
5		5		5	04	5	38
6		6		6		6	
7		7		7	4	7	5
8		8	2	8		8	
9	12	9		9		9	
10	8	10		10		10	
11		11	113	11		11	
12		12		12		12	
13	5	13		13		13	

Minitab output:

```
ANALYSIS OF VARIANCE

SOURCE    DF       SS       MS        F        p
FACTOR     3   100.65    33.55    12.08    0.001
ERROR     12    33.33     2.78
TOTAL     15   133.97
```

```
                                 INDIVIDUAL 95 PCT CI'S FOR MEAN
                                 BASED ON POOLED STDEV
LEVEL    N     MEAN    STDEV   ------+---------+---------+---------+
0        4   10.650    2.053                        (-------*------)
1000     4   10.425    1.486                        (-------*------)
5000     4    5.600    1.244      (------*-------)
10000    4    5.450    1.771      (------*------)
                                 ------+---------+---------+---------+
POOLED STDEV =    1.667            5.0       7.5      10.0      12.5
```

9.17 The stemplots (below) show that the Jaguar XJ12 and the Rolls-Royce Silver Spur are again low outliers, so we omit them. The Mercedes-Benz S420 and S500 are also somewhat low among large cars. However, they are not as extreme as the other two, so one might decide to keep them in.

Descriptive statistics follow the stemplots, appearing as slightly modified Minitab input (some of the statistics supplied by Minitab are not needed here). The variable "Large" includes the two Mercedes-Benz cars, while "Large2" does not. Finally, there are two ANOVAs: the first uses "Large"; the second uses "Large2."

In the first ANOVA, the test is significant at the 1% level; in the second, it is significant at the 5% level. Both sets of confidence intervals show some overlap, but suggest a conclusion similar to that for city mileage—the most important difference is that compact cars have better average mileage than the other two types.

	Compact		Midsize		Large
15		15	0	15	
16	0	16		16	
17		17		17	
18		18		18	
19		19		19	0
20		20		20	0
21		21		21	
22		22	0	22	
23		23	000	23	
24	0000	24	0	24	0
25	00	25	00	25	000
26	000	26	0000	26	000
27	000	27	00	27	00
28	0	28	0	28	00
29	00000	29	000	29	
30	0	30		30	
31	0	31	00	31	
32	00	32		32	
33	0	33		33	
34		34		34	
35	00	35		35	

Minitab output:

	N	MEAN	MEDIAN	STDEV	MIN	MAX	Q1	Q3
Compact	25	28.240	28.000	3.345	24.000	35.000	25.500	30.500
Midsize	19	26.316	26.000	2.689	22.000	31.000	24.000	29.000
Large	13	25.077	26.000	2.753	19.000	28.000	24.500	27.000
Large2	11	26.091	26.000	1.300	24.000	28.000	25.000	27.000

ANALYSIS OF VARIANCE

SOURCE	DF	SS	MS	F	p
FACTOR	2	94.55	47.28	5.21	0.009
ERROR	54	489.59	9.07		
TOTAL	56	584.14			

```
                                  INDIVIDUAL 95 PCT CI'S FOR MEAN
                                  BASED ON POOLED STDEV
    LEVEL      N      MEAN    STDEV    ---+---------+---------+---------+---
    Compact   25    28.240    3.345                       (-----*-----)
    Midsize   19    26.316    2.689             (------*------)
    Large     13    25.077    2.753   (--------*--------)
                                       ---+---------+---------+---------+---
    POOLED STDEV =     3.011          24.0      26.0      28.0      30.0
```

```
ANALYSIS OF VARIANCE
SOURCE      DF        SS        MS        F        p
FACTOR       2     55.26     27.63     3.46     0.039
ERROR       52    415.57      7.99
TOTAL       54    470.84
                                    INDIVIDUAL 95 PCT CI'S FOR MEAN
                                    BASED ON POOLED STDEV
LEVEL        N      MEAN     STDEV    --------+---------+---------+--------
Compact     25    28.240     3.345                          (------*-------)
Midsize     19    26.316     2.689           (-------*--------)
Large2      11    26.091     1.300         (----------*----------)
                                    --------+---------+---------+--------
POOLED STDEV =     2.827              25.5      27.0      28.5
```

9.18 **(a)** $t = -0.34135$ with 12 d.f.; $P = 0.7387$. **(b)** $\overline{x} = 2.344$, MSG $= 0.02652$, MSE $= 0.25411$, and $F = 0.10436$; $P = 0.7491$. **(c)** The two P-values differ by about 0.01—an unimportant difference in most cases.

9.19 Using the means and standard deviations from 9.16(a): $\overline{x} = \frac{1}{16}\big[(4)(10.650) + (4)(10.425) + (4)(5.600) + (4)(5.450)\big] = 8.031$; all other values can be confirmed from the Minitab output in 9.16. Table D places the P-value at less than 0.001; software gives $P = 0.0006$.

CHAPTER 10 SOLUTIONS

10.1 (a) See exercise 2.11 for scatterplot. $r = 0.99415$ and $\hat{y} = -3.660 + 1.19690x$. The scatterplot shows a strong linear relationship, which is confirmed by r. **(b)** β represents how much increase we can expect in humerus length when femur length increases by 1 cm. b (the estimate of β) is 1.1969; $a = -3.660$. **(c)** The residuals are -0.82262, -0.36682, 3.04248, -0.94202, and -0.91102; the sum is 0. $s = \sqrt{3.92843} = 1.9820$.

10.2 (a) See exercise 2.33 for scatterplot. $r = 0.99899$ and $\hat{y} = 1.76608 + 0.080284x$. The scatterplot shows a strong linear relationship; steps per second seem to increase steadily with speed. **(b)** The residuals are 0.0106220, -0.0012674, -0.0010433, -0.0109613, -0.0093443, 0.0031464, and 0.0088482; the sum is 0.0000003. **(c)** $a = 1.766080$, $b = 0.080284$, and $s = \sqrt{0.000082236} = 0.0090684$.

10.3 Using a $t(14)$ distribution: $0.188999 \pm (1.761)(0.004934) = 0.1803$ to 0.1977.

10.4 Using a $t(10)$ distribution: $0.687747 \pm (2.228)(0.2300) = 0.1753$ to 1.2002. β is the increase in second-round score we expect based on an increase of one shot in round one.

10.5 (a) $\hat{y} = -3.6596 + 1.1969$. **(b)** $t = \dfrac{1.1969}{0.0751} = 15.94$. **(c)** 3 d.f.; $P < 0.001$.

10.6 (a) r^2 is very close to 1, which means that nearly all the variation in steps per second is accounted for by foot speed. Also, the P-value for β is small. **(b)** β (the slope) is this rate; the estimate is listed as the coefficient of 'Speed': 0.080284. Using a $t(5)$ distribution: $0.080284 \pm (4.032)(0.0016) = 0.07383$ to 0.08674.

10.7 (a) The plot shows a strong positive linear relationship. **(b)** β (the slope) is this rate; the estimate is listed as the coefficient of 'year': 9.31868. **(c)** 11 d.f.; $t^* = 2.201$; $9.31868 \pm (2.201)(0.3099) = 8.6366$ to 10.0008.

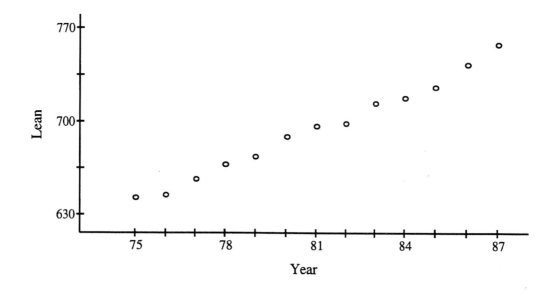

10.8 (a) $\hat{y} = 8.65$ hundred cubic feet per day; savings: 115 ("1.15 hundred") ft^3/day. **(b)** Use the prediction interval: 7.8767 to 9.4217. **(c)** Use the confidence interval: 8.3882 to 8.9101.

10.9 (a) See exercise 2.4 for plot; Powerboat registrations is explanatory. **(b)** The plot shows a moderately strong positive linear relationship; there are no clear outliers or strongly influential points. **(c)** $r^2 = 88.6\%$ indicates that much, but not all, of the variation in manatee deaths is explained by powerboat registrations. **(d)** β is the number of additional manatee deaths we can expect when there are 1000 additional powerboat registrations. Using a $t(12)$ distribution: $0.12486 \pm (1.782)(0.01290) = 0.1019$ to 0.1478. **(e)** $\hat{y} = 45.972$ (about 46 manatee deaths per year). **(f)** Use the confidence interval: 41.49 to 50.46.

10.10 Use $t^* = 1.782$ from part (d), and $SE_{\hat{\mu}} = 2.06$: $45.97 \pm (1.782)(2.06) = 42.30$ to 49.64.

10.11 (a) The stemplot does not show any *major* asymmetry, and also has no particular outliers. **(b)** The plot does not suggest a nonlinear relationship. There is *some* indication that there may be less variation at the high and low ends of the plot, but nothing too strong—there are too few observations to make any judgments about that.

```
-0 | 7
-0 |
-0 | 33
-0 | 110
 0 | 0011111
 0 | 2
 0 | 4
 0 | 6
```

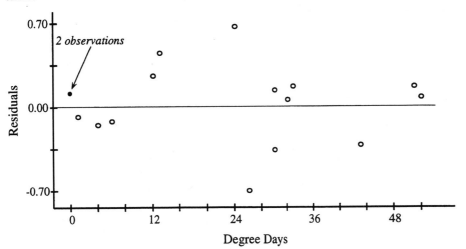

10.12 The number of points is so small that it is hard to judge much from the stemplot. The scatterplot of residuals vs. year does not suggest any problems. The regression in 10.7 should be fairly reliable.

```
-0 | 6
-0 | 55
-0 | 32
-0 |
 0 | 011
 0 | 22
 0 | 44
 0 | 7
```

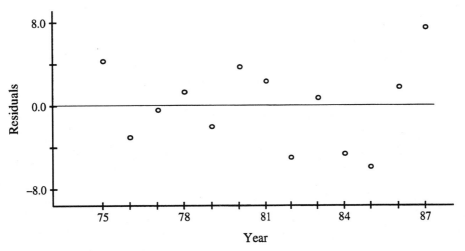

Chapter Review

10.13 (a) See exercise 2.42 for scatterplot. **(b)** $r^2 = 0.324$; $r = \sqrt{0.324} = 0.569$—use the *positive* square root since the relationship is clearly positive, and r must have the same sign as b. (In fact, $r = 0.56896$.) The regression of overseas returns on US returns explains about 1/3 (32.4%) of the variation in overseas returns; the relationship is positive. There is one outlier (in 1986) and one potentially influential point (in 1974). **(c)** $H_0 : \beta = 0$ vs. $H_a : \beta \neq 0$; $t = 3.09$, and $P = 0.006$ (so we reject H_0). **(d)** $\hat{y} = 21.037$ percent. **(e)** Use the prediction interval: −21.97 to 64.04 percent. In practice, this is of little value—the interval includes everything from a 20% loss to a 60% gain.

10.14 Use $t^* = 2.977$ from a $t(14)$ distribution: $8.6492 \pm (2.977)(0.1216) = 8.287$ to 9.011.

10.15 (a) It appears that the variation about the line is greater for larger values of x—on the left side of the plot, the residuals are less spread out. **(b)** Stemplot on right (residuals were rounded to whole numbers first).

```
-2 | 54310
-1 | 75
-0 | 88732
 0 | 01
 1 | 03669
 2 | 36
 3 |
 4 |
 5 | 0
```

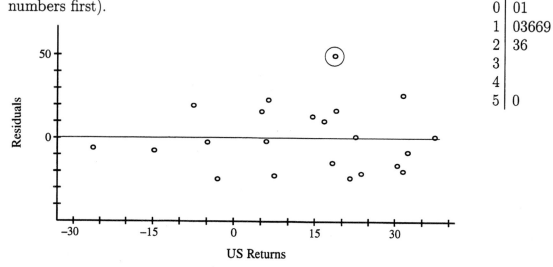

10.16 (a) and **(b)** See exercise 2.86 for the plot; the influential points are the three in the upper right. **(c)** $H_0 : \beta = 0$ vs. $H_a : \beta \neq 0$. $t = 4.28$, df $= 15$, $P = 0.001$ (actually, $P < 0.001$). (Based on the statement, it might be reasonable to say that H_a is $\beta > 0$, a one-sided alternative, which would only make the P-value half as large.) **(d)** Franklin's predicted income was $\hat{y} = 21.6$ million dollars; the residual is −7.8. **(e)** Use the prediction interval: −30.96 to 74.17. Franklin's actual income seems quite reasonable when compared to this wide interval; we would not have been able to predict Franklin's problems from this.

10.17 (a) $\hat{y} = 26.3320 + (0.687747)(89) = 87.5415$ (using the output of Figure 10.5), residual $= 6.4585$. **(b)** $\sum \text{residual}^2 = 356.885$; $s = \sqrt{\frac{1}{10}(356.885)} = \sqrt{35.6885} = 5.974$.

10.18 (a) $r = \sqrt{\text{"R squared"}} = \sqrt{0.472} = 0.687$—use the *positive* square root since r must have the same sign as b. **(b)** $P = 0.0136$ (taken from Figure 10.5). **(c)** $P = \frac{1}{2}(0.0136) = 0.0068$.

10.19 Take $t^* = 2.145$ from a $t(14)$ distribution: $1.0892 \pm (2.145)(0.1389) = 0.7913$ to 1.3871.

10.20 (a) $\bar{x} = 89.67$ and $s_x = 7.83$. (b) $x^* = 90$; $(x^* - \bar{x})^2 = 0.1089$. (c) $\sum(x - \bar{x})^2 = (11)(7.83)^2 = 674.4$. (d) $SE_{\hat{\mu}} = (5.974)\sqrt{\frac{1}{12} + \frac{0.1089}{674.4}} = 1.7262$. (e) $\hat{y} = 88.23$; confidence interval: $88.23 \pm (2.228)(1.7262) = 84.38$ to 92.08.

10.21 (a) City mileage is the explanatory variable. NOTE: Numbers (rather than circles) mean that there were several cars with the same mileage pair. (b) There is a fairly strong positive linear association. (c) $\hat{y} = 6.178 + 1.03764x$ (the solid line). (d) After removing the two circled points: $\hat{y}^* = 7.745 + 0.96251x$ (the dashed line). Original s was 1.631; without Jaguar and Rolls-Royce, $s = 1.592$. (e) The Toyota Corolla has the largest absolute residual at -4.157—the predicted value is higher than the actual value.

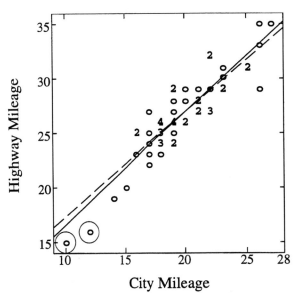

City Mileage

(f) The stemplot shows only the mild low outlier for the Corolla; it has no other disturbing departures from normality. Plotting residuals vs. the x values (city mileage) shows a slight problem in that all the residuals for low x values are negative. (g) β is the average increase in highway mileage for each increase of 1 unit in city mileage. Confidence interval (using 50 d.f.): $1.03764 \pm (2.009)(0.06476) = 0.9075$ to 1.1677. (h) $\hat{y} = 24.856$; we want a prediction interval, which software gives as 21.555 to 28.157.

```
-4 | 2
-3 |
-2 | 6000
-1 | 9998776110000
-0 | 9999998200
 0 | 001111111228
 1 | 0011111128
 2 | 1122
 3 | 00112
```

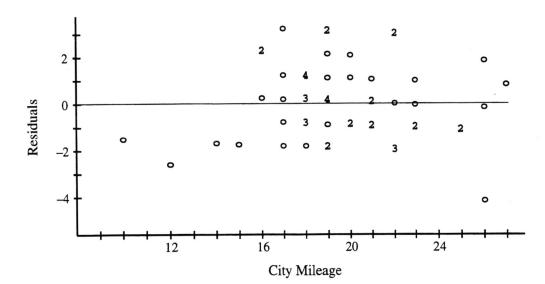

City Mileage

10.22 (a) The heavy fish does not appear to be out of place on the width vs. length plot (below). (However, when Minitab does the regression for (b), it notes that this fish has the largest residual.) **(b)** $\hat{y} = -0.8831 + 0.297518x$. **(c)** Based on Minitab output: $\hat{y} = 6.5549$; confidence interval is $6.5549 \pm 0.1294 = 6.4255$ to 6.6842. This is based on $SE_{\hat{\mu}} = 0.0645$ and $t^* = 2.005$ from a $t(54)$ distribution. **(d)** A stemplot of the residuals (right) gives *some* suggestion of right-skewness, and has two moderate low outliers. A plot of residuals vs. length suggests that there may be more variability in width for larger lengths, but that may just be because we have few observations for small fish. There are no apparent gross violations, so inference should be fairly safe.

```
-1 | 0
-0 | 8
-0 |
-0 | 555544444
-0 | 3332222222
-0 | 111110000
 0 | 000000111
 0 | 22333
 0 | 444555
 0 | 666
 0 | 8
 1 | 0
 1 | 2
```

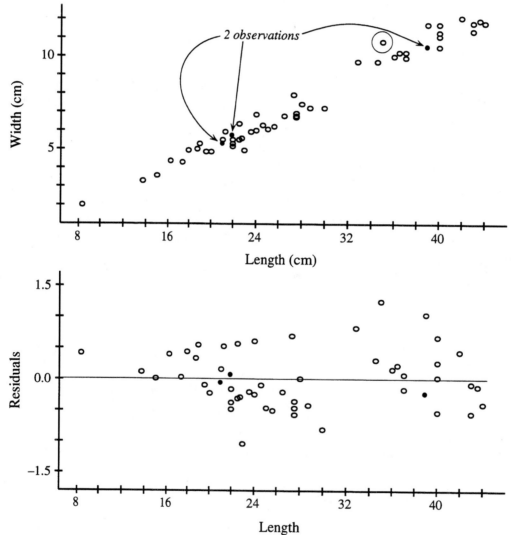

10.23 (a) The plot shows only a *very* slight pattern (weight increases with length). There is a definite outlier: fish #40, the "heavy fish" mentioned in the previous problem.

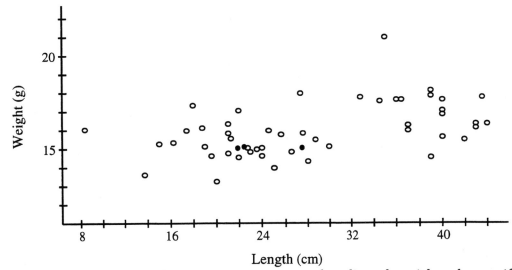

Length (cm)

(b) We would expect weight to increase more-or-less linearly with volume—if we double volume, we double weight; if we triple the volume, we triple the weight, etc. When all dimensions (length, width, and height) are doubled, the *volume* of an object increases by a factor of $8 = 2^3$. Similarly, if we triple all dimensions, volume (and, approximately, weight) increases by a factor of $27 = 3^3$. It then makes sense that the cube root (i.e., the one-third power) of the weight increases at an approximately linear rate with length. **(c)** The second plot omits the heavy fish—which would still be an outlier. It does not really look any more linear; there is still a lot a scatter that obscures any obvious patterns. There are no apparent outliers (except for the heavy fish).

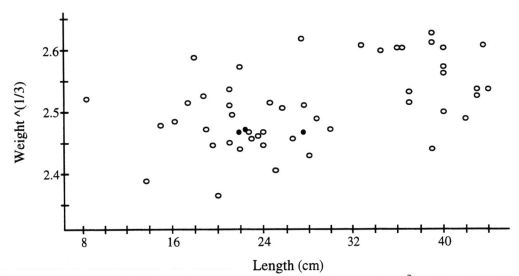

Length (cm)

(d) There is very little change in appearance of the plots, or in r^2: using the original weight variable, $r^2 = 0.2346$; with weight$^{1/3}$, $r^2 = 0.2389$. In fact, if we omit the outlier from both computations, we actually find that r^2 *decreases*: 0.2515 before, 0.2502 after. **(e)** If we use all the fish: $\hat{y} = 2.40306 + 0.0038135x$; $\hat{y} = 2.49839$ when

$x = 25$; confidence interval: $2.49839 \pm t^* SE_{\hat{\mu}} = 2.49839 \pm (2.005)(0.00870) = 2.48094$ to 2.51585.

Without fish #40: $\hat{y} = 2.40899 + 0.0034570x$; $\hat{y} = 2.49541$ when $x = 25$; confidence interval: $2.49541 \pm (2.006)(0.00772) = 2.47994$ to 2.51089. **(f)** Both the stemplot and the plot of residuals vs. length show no gross violations of the assumptions, except for the high outlier for fish #40; as we saw in part (e), however, omitting that fish makes little difference in our line.

```
-1 | 11
-0 | 987655
-0 | 44444433333322222221110
 0 | 000011112344
 0 | 5555667788
 1 | 01
 1 |
 2 | 1
```

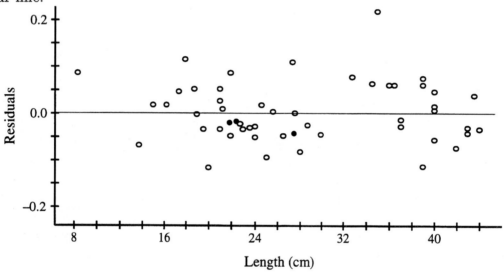